# Gas Turbines Modeling, Simulation, and Control

## Using Artificial Neural Networks

# Gas Turbines Modeling, Simulation, and Control
## Using Artificial Neural Networks

Hamid Asgari
XiaoQi Chen

**CRC Press**
Taylor & Francis Group
Boca Raton  London  New York

CRC Press is an imprint of the
Taylor & Francis Group, an **informa** business

CRC Press
Taylor & Francis Group
6000 Broken Sound Parkway NW, Suite 300
Boca Raton, FL 33487-2742

First issued in paperback 2017

© 2016 by Taylor & Francis Group, LLC
CRC Press is an imprint of Taylor & Francis Group, an Informa business

No claim to original U.S. Government works

ISBN-13: 978-1-4987-2661-0 (hbk)
ISBN-13: 978-1-138-89344-3 (pbk)

# Contents

# List of figures

# List of tables

# Preface

Gas turbines (GT) are one of the significant parts of modern industry. They play a key role in the aeronautical industry, power generation, and main mechanical drivers for large pumps and compressors. Modeling and simulation of GTs has always been a powerful tool for the performance optimization of this kind of equipment. Remarkable activities have been carried out in this field and a number of analytical and experimental models have been built so far to get in-depth understanding of the nonlinear behavior and complex dynamics of these systems. However, the need to develop accurate and reliable models of GTs for different objectives and applications has been a strong motivation for researchers to continue to work in this fascinating area. The study in this field includes white-box- and black-box-based models and their applications in control systems. Artificial neural networks (ANNs) as a black-box methodology have been regarded as suitable and powerful tools for data processing, modeling, and control of highly nonlinear systems such as GTs. Besides, because of the high demand of the electricity market, the power producers are eager to continuously investigate new methods of optimization for design, manufacturing, control, and maintenance of GTs. In recent decades, ANNs have shown a high and strong potential to be considered as a reliable alternative to the conventional modeling, simulation, and control methodologies.

This book presents novel methodologies for modeling, simulation, and control of GTs using ANNs. In the field of modeling and simulation, two different types of GTs are modeled and simulated using both Simulink® and neural network-based models. Simulated and operational data sets are employed to demonstrate the capability of neural networks in capturing complex nonlinear dynamics of GTs. For ANN-based modeling, the applications of both static (MLP) and dynamic (NARX) networks are explored. Simulink and NARX models are set up to explore both steady-state and transient behaviors. The models developed in this book can be used offline for design and manufacturing purposes or online on sites for condition monitoring, fault detection, and troubleshooting of GTs. This book provides new approaches and novel solutions to modeling, simulation, and control of GTs. It is structured as follows:

**Chapter 1** discusses the motivations, necessity, and goal of modeling and control of GTs. A classification of GTs is provided and main considerations in GT modeling are presented. The chapter briefly explains the most important criteria and considerations at the beginning of the GT modeling process including GT types and configurations, modeling methods, control system types and configurations, as well as modeling objectives and approaches. The chapter also defines the current problems in the area of modeling, simulation, and control of GTs. Finally, it briefly provides the scope and main objectives of current and future activities in the area of GT modeling and simulation.

**Chapter 2** presents a comprehensive overview of the research activities in the field of white-box modeling, simulation, and control of GTs based on the classification of GTs. The most relevant scientific sources for different kinds of GTs including low-power, industrial power plant, and aero GTs are explored in this chapter.

**Chapter 3** gives a comprehensive overview of the most significant studies in the field of black-box modeling, simulation, and control of GTs based on the classification of GTs. It covers models of low-power GTs, industrial power plant gas turbine (IPGT), and aero GTs.

**Chapter 4** briefly discusses the structure of ANNs and ANN-based model building processes, including system analysis, data acquisition and preparation, network architecture, as well as network training and validation. It explores different challenges in using ANN-based models for industrial systems and describes the advantages and limitations of this approach.

**Chapter 5** introduces a novel ANN-based methodology for offline system identification of a low-power single-shaft GT. The processed data is obtained from a SIMULINK model of a GT in MATLAB® environment. A comprehensive computer program code is generated and run in MATLAB for creating and training different ANN models with feedforward multilayer perceptron (MLP) structure. The code consists of 18,720 different ANN structures including various training functions, different number of neurons, as well as a variety of transfer (activation) functions for hidden and output layers of the network.

**Chapter 6** presents modeling of the transient behavior of GTs. Simulink and NARX models are created and validated using experimental data sets to explore transient behavior of a heavy-duty IPGT. The results show that both Simulink and NARX models successfully capture the dynamics of the system. However, a NARX approach can model GT behavior with a higher accuracy compared to a Simulink approach. Besides, a separate complex model of the start-up operation of the same IPGT is built and verified by using NARX models. The models are set up and verified on the basis of measured time-series data sets. It is observed that NARX models have the potential to simulate start-up operation and to predict dynamic behavior of GTs.

**Chapter 7** gives a model of the start-up operation of a heavy-duty IPGT by using NARX models by using the data taken experimentally during the start-up procedure. The NARX model is set up on the basis of three measured time-series data sets for two different maneuvers. To verify the resulting models, they are applied to three other available data sets and comparisons are made among significant outputs of the models and the values of the corresponding measured data.

**Chapter 8** elucidates a neural network approach for controller design of GTs. A conventional proportional-integral-derivative (PID) controller and neural network-based controllers consisting of ANN-based model predictive (MPC) and feedback linearization (NARMA-L2) controllers are designed and employed to control rotational speed of a GT. The related parameters for all controllers are tuned and set up according to the requirements of the controllers design. It is demonstrated that neural network-based controllers (in this case NARMA-L2) can perform even better than conventional controllers. The settling time, rise time, and maximum overshoot for the response of NARMA-L2 is less than the corresponding factors for the conventional PID controller. It also follows the input changes more accurately than the PID.

This book can be an invaluable source of research for graduate and postgraduate students, researchers, mechanical, mechatronics, and control engineers, as well as GT manufacturers and professionals who deal with artificial intelligence, neural network, GTs, and industrial equipment. Readers can learn how artificial intelligence can be used to solve complicated industrial problems specifically in the area of GTs. This book can also be used as a rich source of information about research activities in the field of modeling, simulation, and control of GTs.

MATLAB® and Simulink® are registered trademarks of The MathWorks, Inc. For product information, please contact:

The MathWorks, Inc.
3 Apple Hill Drive
Natick, MA 01760-2098 USA
Tel: 508 647 7000
Fax: 508-647-7001
E-mail: info@mathworks.com
Web: www.mathworks.com

# Acknowledgments

The authors would like to express their sincere appreciation to Dr. Raazesh Sainudiin in the Department of Mathematics and Statistics at the University of Canterbury (UC) and Professor Mohammad Bagher Menhaj in the Electrical Engineering Department of the Amir Kabir University of Technology (AUT) in Iran who have greatly assisted in the research resulting in the writing this book. We would also like to express our gratitude to Dr. Mark Jermy and Dr. Sid Becker, the postgraduate coordinators in the Mechanical Engineering Department, Professor Milo Kral, the head of the Mechanical Engineering Department, and Professor Lucy Johnston, the dean of Postgraduate Research at UC, for their help and support.

Our appreciation also goes to Professor Pier Ruggero Spina, Professor Mauro Venturini, Professor Michele Pinelli, and Dr. Mirko Morini in "Dipartimento di Ingegneria" at "Università degli Studi di Ferrara" in Italy for providing experimental data and their close collaboration in development of Simulink and ANN models. We also warmly thank Assistant Professor Mohsen Fathi Jegarkandi at the Aerospace Department of Sharif University of Technology and Dr. Naser Rahbar at Malek-Ashtar University of Technology (MUT) who have greatly helped the authors for development of the control system of GTs. We would like to express our sincere gratitude to the staff at Iranian Offshore Oil Company (IOOC) for their technical support and close cooperation during site visit and data acquisition.

It is our great pleasure to acknowledge the help and support of the UC staff in Admission and Enrolment, ICT Services, Recreation Centre, Student Services, and UC libraries. We are also very grateful to all technicians, administrative staff, fellow postgraduates, students, and friends at UC whom we had the pleasure of working and cooperating with.

# *Authors*

**Hamid Asgari** received his PhD in mechanical engineering from the University of Canterbury (UC) in Christchurch, New Zealand in 2014. He obtained his MSc in aerospace engineering from Tarbiat Modares University (TMU), Tehran, Iran, and his BEng in mechanical engineering from Iran University of Science & Technology (IUST), Tehran, Iran. He has worked more than 15 years in his professional field as a lead mechanical engineer and project coordinator in very highly-prestigious industrial companies. During his professional experience, he has been a key member of engineering teams in design, R&D, and maintenance planning departments. He has invaluable theoretical and hands-on experience in technical support, design, and maintenance of a variety of mechanical equipment and rotating machinery, such as GTs, pumps, and compressors, in large-scale projects in power plants and the oil and gas industry.

**XiaoQi Chen** is a professor in the Department of Mechanical Engineering at the University of Canterbury, Christchurch, New Zealand. After obtaining his BEng from South China University of Technology, Guangzhou, China, in 1984, he received the China-UK Technical Co-Operation Award for his MSc study in the Department of Materials Technology, Brunel University, Uxbridge, London, UK (1985–1986), and PhD study in the Department of Electrical Engineering and Electronics, University of Liverpool, Liverpool, Merseyside, UK (1986–1989). He was a senior scientist at the Singapore Institute of Manufacturing Technology (1992–2006) and a recipient of the Singapore National Technology Award in 1999. His research interests include mechatronic systems, mobile robotics, assistive devices, and manufacturing automation.

# Nomenclature

## Abbreviations

| | |
|---|---|
| AMPC | Approximate model predictive control |
| ANFIS | Adaptive network-based fuzzy inference system |
| ANN | Artificial neural network |
| ARX | Autoregressive with exogenous input |
| ASME | American Society of Mechanical Engineers |
| BPNN | Backpropagation neural network |
| CCGT | Combined cycle gas turbine |
| CCPP | Combined cycle power plant |
| CO | Carbon monoxide |
| CSGT | Control system of gas turbine |
| CT | Compressor turbine |
| CUSMUS | Cumulative sum (technique) |
| DC | Direct current |
| DCS | Distributed control system |
| DLE | Dry low emission |
| DLN | Dry low nitrogen oxide |
| DNN | Dynamic neural network |
| FDI | Fault detection and isolation |
| FFNN | Feedforward neural network |
| GA | Genetic algorithm |
| GAST | Gas turbine governor model |
| GG | Gas generator |
| GPC | Generalized predictive control |
| GT | Gas turbine |
| HDGT | Heavy-duty gas turbine |
| HP | High pressure (gas turbine) |
| IGV | Inlet guide vane |
| IPGT | Industrial power plant gas turbine |
| LP | Low pressure (gas turbine) |
| MGT | Microgas turbine |

| MIMO | Multiple-input and multiple-output |
|------|-----------------------------------|
| MLP | Multilayer perceptron |
| MP | Minimum phase |
| MPC | Model predictive control |
| MSE | Mean square error |
| NARMA | Nonlinear autoregressive moving average |
| NARMA-L2 | Feedback linearization control |
| NARMAX | Nonlinear autoregressive moving average with exogenous inputs |
| NARX | Nonlinear autoregressive with exogenous inputs |
| NMP | Nonminimum phase |
| NMPC | Nonlinear model predictive control |
| NN | Neural network |
| NO | Nitrogen oxide |
| OEM | Original equipment manufacturer |
| PD | Proportional-derivative (controller) |
| PI | Proportional-integral (controller) |
| PID | Proportional-integral-derivative (controller) |
| PR | Pressure ratio |
| PT | Power turbine |
| RBF | Radial basis function |
| RBFNN | Radial basis function neural network |
| RL | Reinforcement learning |
| RMSE | Root mean square error |
| RNN | Recurrent neural network |
| SIMO | Single-input and multiple-output |
| SISO | Single-input and single-output |
| TDL | Time delay |
| TIT | Turbine inlet temperature |
| TOT | Turbine outlet temperature |
| UPFC | Unified power flow controller |
| VSV | Variable stator vane |

## *Variables*

| | |
|------|------|
| $C_p$ | Specific heat in constant pressure (J/kg K) |
| $C_v$ | Specific heat in constant volume (J/kg K) |
| $HR$ | Heat rate (kJ/kWh) |
| $I$ | Moment of inertia (kg $\cdot$ m$^2$) |
| $J$ | Cost function |
| $LHV$ | Lower heating value of fuel (J/kg) |
| $m$ | Mass (kg) |
| $\dot{m}$ | Mass flow rate (kg/s) |
| $M$ | Momentum (N m) |

| $N$ | Rotational speed (rpm, or 1/s) |
| $P$ | Stagnation pressure (Pa) |
| $q$ | Lower thermal value (J/kg) |
| $Q$ | Heat power (W) |
| $R$ | Specific gas constant (J/kg K) |
| $S$ | Entropy (J/K) |
| *SFC* | Specific fuel consumption (kg/kWh) |
| $t$ | Time (s) |
| $T$ | Temperature (K) |
| $U$ | Externally determined variable (system input) |
| $u'$ | Tentative control signal |
| $V$ | Volume (m$^3$) |
| $W$ | Work (J) |
| $\dot{W}$ | Power (W) |
| $Y$ | Variable of interest (system output) |

## Constants

| $C$ | Pressure constant |
| $D$ | Delay |
| $F$ | Fuel to air mass flow rates ratio |
| $J$ | A natural number |
| $K$ | A natural number |
| $N$ | A natural number |
| $N_1, N_2, N_3, ..., N_u$ | Horizons (MPC factors) |
| $PR_C$ | Compressor pressure ratio |

## Subscripts

| 00 | Ambient |
| 01 | Compressor inlet |
| 02 | Compressor outlet |
| 03 | Turbine inlet |
| 04 | Turbine outlet |
| A | Average value (for compression process in compressor) |
| C | Compressor |
| Cc | Combustion chamber |
| D | Data set |
| F | Fuel |
| G | Average value (for expansion process in turbine) |
| Gt | Gas turbine |
| In | Inlet |
| M | Measured |
| Mech | Mechanical |

| Med | Medium |
| Out | Outlet |
| R | Reference (desired) |
| T | Turbine |
| U | System input |
| Y | System output |

## General symbols

| D | Derivative (controller) |
| E | Error |
| F | Function |
| I | Integral (controller) |
| M | Maneuver |
| N | Number |
| P | Proportional (controller) |
| Trainbfg | BFGS Quasi-Newton training algorithm |
| Trainbr | Bayesian regularization training algorithm |
| Traincgb | Conjugate gradient with Powell/Beale restarts training algorithm |
| Traincgf | Fletcher-Powell conjugate gradient training algorithm |
| Traincgp | Polak-Ribiére conjugate gradient training algorithm |
| Traingdx | Variable learning rate gradient descent training algorithm |
| Trainlm | Levenberg-Marquardt training algorithm |
| Trainoss | One step secant training algorithm |
| Trainrp | Resilient backpropagation training algorithm |
| Trainscg | Scaled conjugate gradient training algorithm |

## Greek symbols

| $\gamma$ | Ratio of specific heats |
| $\eta$ | Efficiency |
| $\xi$ | Pressure loss coefficient |
| $\rho$ | Contribution |

# chapter one

# Introduction to modeling of gas turbines

> The knowledge of anything, since all things have causes, is not acquired or complete unless it is known by its causes.

> **Ibn Sina (Avicenna)**
> *Persian Polymath, 980–1037*

A gas turbine (GT) is an internal combustion engine that uses the gaseous energy of air to convert chemical energy of fuel into mechanical energy. It is designed to extract, as much as possible, the energy from the fuel [1]. The service of GTs in industrial equipment and utilities located on power plants and offshore platforms has increased in the past 50 years. This high demand is because of their low weight, compactness, and multiple fuel applications [2]. Although the story of GTs has taken root in history, it was not until 1930s that the first practical GT was developed by Frank Whittle and his colleagues in Britain for a jet aircraft engine [3]. GTs were developed rapidly after World War II and became the primary choice for many applications. That was especially because of the enhancement in different areas of science such as aerodynamics, cooling systems, and high-temperature materials, which significantly improved the engine efficiency. Thus, it is not surprising that GTs have been increasing in popularity year by year. They have the ability to provide a reliable and continuous operation. The operation of nearly all available mechanical and electrical equipment and machinery in industrial plants such as petrochemical plants, oil field platforms, gas stations, and refineries, depends on the power produced by GTs. The wide application of GTs throughout the world especially in electrical utilities is due to their reliability, availability, adaptability, fast start capability, low initial cost, and short delivery time [4]. They are independent of cooling water and can operate on a variety of fuels. GTs provide high rates of load growth in the summer time and respond fast to load changes [4].

During recent years, considerable activities have been carried out especially in the field of modeling and simulation of GTs. It is because the need for and use of GTs have become more apparent in the modern industry. Creating models of GTs and their related control systems has been a useful technical and cost-saving strategy for performance optimization of the equipment before the final design process and manufacturing. GT models and simulators can be used for off-design performance prediction, and evaluation of emissions, turbine creep life usage, and the engine control system [5]. Mathematical modeling is considered as a general methodology for system modeling. It uses mathematical language to describe and predict the behavior of a system.

Great efforts have been made in developing GTs to overcome their related challenging economic and engineering problems [6] and to have a reliable and cost-effective design [7]. One of the best approaches for optimization of design, performance, and maintenance of GTs is offline modeling and simulation. It helps manufacturers and users in different ways. Manufacturers can evaluate and optimize the performance of a specific model of GT during design and manufacturing processes. Models may also be used online on sites by operators and site engineers for condition monitoring, sensor validation, fault detection, troubleshooting, and so on. A variety of analytical and experimental models of GTs has been built so far. However, the need for optimized models for different objectives and applications has been a strong motivation for researchers to continue to work in this area.

This chapter briefly explains the principles of performance of a typical GT and classification of GTs. Main considerations in GT modeling including GT types and configurations, modeling methods, control system types and configurations, as well as modeling objectives and approaches, are explained in the following sections of this chapter; followed by problems and limitations and modeling objectives.

## 1.1   GT performance

GTs work based on Brayton cycle: Figure 1.1 shows a typical single-shaft GT and its main components including compressor, combustion chamber

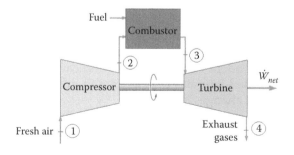

*Figure 1.1* (**See color insert.**) Schematic of a typical single-shaft GT.

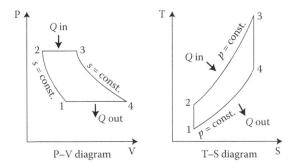

***Figure 1.2*** Typical Brayton cycle in pressure–volume and temperature–entropy frames. (From Wikimedia Commons Webpage. 2013. [Online]. Available: http://en.wikipedia.org. [Accessed June 14, 2013].)

(combustor), and turbine. The set of these components is called engine core or gas generator (GG). Compressor and turbine are connected to the central shaft, and they rotate together.

Figure 1.2 shows standard Brayton cycle in pressure–volume (P–V) and temperature–entropy (T–S) frames, respectively [8]. As the figure shows, air enters the compressor at section 1 and is compressed through passing the compressor. The hot and compressed air enters the combustion chamber (combustor) at section 2. In combustor, fuel is mixed with air and ignited. The hot gases which are the product of combustion are forced into the turbine at section 3 and rotate it. Turbine drives the compressor and the GG mechanical output, which can be an electricity alternator in a power plant station, a large pump, or a large compressor. The ideal processes in the compressor (1-2) and turbine (3-4) are isentropic. There is also an isobaric process in the combustor (2-3) and environment (4-1) for the ideal cycle. However, the actual processes in the compressor and turbine are irreversible and nonisentropic. There is also a pressure loss during the process in the combustor. Neglecting the pressure loss in the air filters and the combustor, processes 2-3 and 4-1 can be considered isobaric [9].

## 1.2   GT classification

GTs can be divided into two main categories including aero GTs and stationary GTs. In aero industry, GT is used as propulsion system to make thrust and move an airplane through the air. Thrust is usually generated based on the Newton's third law of action and reaction. There are varieties of aero GTs including turbojet, turbofan, and turboprop. In stationary GTs, GG may be tied to electro generators, large pumps, or compressors to make turbo-generators, turbo-pumps or turbo-compressors, respectively.

If the main shaft of the GG is connected to an electro generator, it can be used to produce electrical power.

In another classification, GTs can be divided into the following five groups [2], based on their structure, application, and output power (MW):

- Microgas turbines (MGT), with 20–350 KW output power
- Small GTs for simple cycle applications, with 0.5–2.5 MW output power and 15%–25% efficiency
- Aero-derivative GTs for the aerospace industry, with 2.5–50 MW output power and 35%–45% efficiency
- Frame-type heavy-duty gas turbines (HDGT) for large power generation units, with 3–480 MW output power and 30%–46% efficiency
- Industrial-type GTs for extensive use in petrochemical plants, with 2.5–15 MW output power and 30%–39% efficiency

In this book, micro and small GTs are considered as low-power GTs, and industrial types and HDGTs that are used in power plants for generating electricity are called industrial power plant gas turbines (IPGT). IPGTs are playing a key role in producing power, especially for the plants which are far away on oil fields and offshore sites where there is no possibility of connecting to the general electricity network.

## 1.3   Considerations in GT modeling

Before making a GT model, some basic factors should be carefully considered. GT type, GT configuration, modeling methods, control system type and configuration, and modeling objectives are among the most important criteria at the beginning of the modeling process [10].

### 1.3.1   GT type

As the first step of modeling, it is necessary to get enough information about the type of GT, which is to be modeled, as a GT can be an aero or a stationary GT. Although there are different types of GT based on their applications in the industry, they have the same main common parts including compressor, combustion chamber, and turbine. Figure 1.3 shows a typical single-spool aero GT engine [11].

### 1.3.2   GT configuration

Configuration of a GT is another important criterion in GT modeling. Although all GTs nearly have the same basic structure and thermodynamic cycle, there are considerable distinctions when they are investigated in details. For instance, to enhance GT cycle, system efficiency, or

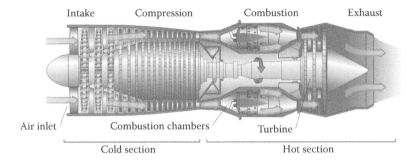

Intake        Compression        Combustion        Exhaust

Air inlet        Combustion chambers        Turbine

Cold section        Hot section

*Figure 1.3* (**See color insert.**) A typical single-spool turbojet engine. (From Wikimedia Commons. 2012. [Online]. Available: http://commons.wikimedia.org.)

output power, through different methods such as reheating, intercooling or heat exchange, particular GT configurations are utilized. GTs can also be categorized based on the type of their shafts. They may be single-shaft or split-shaft (twin-shaft or triple-shaft). In a single-shaft GT, the same turbine rotor, drives the compressor is connected to the power output shaft through a speed reduction. In a split-shaft GT, the GG turbine and the power turbine (PT) are mechanically disconnected. GG turbine, also called compressor turbine (CT) or high-pressure (HP) turbine, is the component, which provides required power for driving the compressor and accessories. However, PT, also called low-pressure (LP) turbine, does the usable work. Figure 1.4 shows a typical twin-shaft GT engine [12].

## 1.3.3   GT modeling methods

Modeling and simulation of GTs play a key role in manufacturing the most efficient, reliable, and durable GTs. Besides, GT models can also be used on industrial sites for optimization, condition monitoring, sensor validation, fault detection, troubleshooting, and so on. These facts have been a strong motivation for scientists to keep carrying out research in this field. There are many sources regarding modeling and control of GTs, and a variety of GT models has been built so far from different perspectives and for different purposes. Although some researchers such as Visser et al. [13] tried to introduce a generic model for GTs using commercial software, the presented models are based on varieties of methodologies and approaches.

Mathematical modeling as a general methodology for system modeling uses mathematical language to describe and predict behavior of a system. Important advances and development of scientific fields may be tied to the quality of mathematical models and their agreement with the results of experimental measurements. Physics-based modeling is a main branch of mathematical modeling. It implies that the system is governed

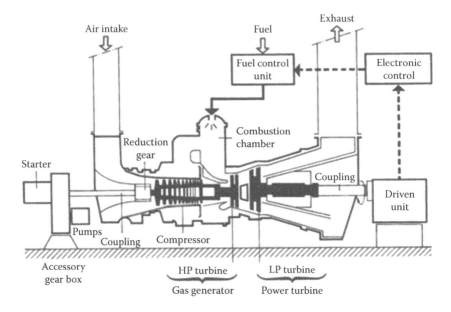

*Figure 1.4* A typical twin-shaft GT engine. (From THM Gas Turbine Basic Training, Turbo Training Catalogues, MAN Diesel & Turbo Co., 2009.)

by the laws of physics, which leads to physically realistic simulation. Physics-based modeling employs equations of mathematical physics, coupled with real-time sensor measurements to set up models suitable for operational usage. Mathematical models can be classified as "linear and nonlinear," "deterministic and stochastic (probabilistic)," "static and dynamic," or "discrete and continuous" [14].

### 1.3.3.1   Linear and nonlinear models

A model is called linear if all objective functions and constraints of the system are represented by linear equations. Otherwise, it is considered as a nonlinear model. Although industrial equipment usually shows nonlinear behavior, in many cases the model is simplified to be analyzed linearly. There are different methods to linearize a nonlinear system. However, in setting up a model, which can accurately predict the behavior of complex and sensitive systems such as GTs, considering nonlinear dynamics is unavoidable.

### 1.3.3.2   Deterministic and stochastic (probabilistic) models

From another perspective, a model can be deterministic or stochastic. In a deterministic model, all variable states are uniquely determined by the parameters in the model and by the sets of previous states of these variables. Therefore, a deterministic model expresses itself without uncertainty due to an exact relationship between measurable and derived variables. Conversely, in a stochastic model, quantities are described using stochastic variables or stochastic processes. Therefore, in a stochastic model, variable states are described using random probability distributions [14].

### 1.3.3.3   Static and dynamic models

The variables that characterize a system usually change with time. If there are direct, instantaneous links among these variables, the system is called static. If the variables of a system change without direct outside influence so that their values depend on earlier applied signals, then the system is called dynamic [14].

### 1.3.3.4   Discrete and continuous models

A mathematical model is called continuous-time when it describes the relationship between continuous-time signals. Continuous-time models are shown with a function f(t) that changes over continuous time. A model is called discrete-time when it directly expresses the relationships between the values of the signals at discrete instants of time. Relationship between signal values is usually expressed by using differential equations. In practical applications, signals are most often obtained in sampled form in discrete-time measurements [14].

## 1.3.4   GT control system type and configuration

One of the most important factors in modeling and control of GTs is type and configuration of their control system. Control system is a vital part of any industrial equipment. Type and configuration of a control system are in a close relationship with the complexity of the system dynamics and the defined tasks during the whole performance period. Lacking a proper control system can lead to serious problems such as compressor surge, overheat, over speed, and so on [15]. The final effect of these problems may be system shutdown and severe damages to the main components of GT.

There are three main functions of the control system of all GTs including "start-up and shutdown sequencing control," "steady-state or operational control," and "protection control for protection from overheat, overspeed, overload, vibration, flameout and loss of lubrication." In a power network with several GTs, all individual control systems are closely connected with a central distributed control system (DCS) [2]. Control system of gas turbines (CSGT) may be open-loop or closed-loop. In an open-loop control system, the manipulated variable is positioned manually or by using a predetermined program. However, to control a device in a closed-loop control system, one or more variables of measured data process parameters are used to move the manipulated variable. To keep the closed-loop control system effective and suitable, the controller should be properly related to the process parameters [2]. Figures 1.5 and 1.6 show open-loop and closed-loop control system block diagrams for a typical process, respectively.

## 1.3.5 GT modeling objectives

There are many different goals for making a model of GTs such as condition monitoring, fault detection and diagnosis, sensor validation, system identification, as well as design and optimization of control system. Thus, a clear statement of the modeling objectives is necessary to make a successful GT model.

### 1.3.5.1 Condition monitoring

One of the objectives of making a GT model condition monitoring. Condition monitoring is considered a major part of preventive maintenance. It assesses the operational health of GTs and indicates potential failure warning(s) in advance, which help operators to take the proper action predicted in preventative maintenance schedule [16]. Condition monitoring is a very helpful tool in maintenance planning and can be used to avoid unexpected failures. Lost production, overtime, and expediting costs can be effectively prevented by predicting failures before any serious damage occurs in the system. To minimize the maintenance costs for very important and expensive machines such as GTs, it is necessary to monitor the operational conditions of vital and sensitive parts of the equipment and to continuously obtain their related data for further analysis. Good condition monitoring reduces the number of wrong decisions, minimizes the demand for spare parts, and reduces maintenance costs. A good maintenance system should be capable of monitoring all

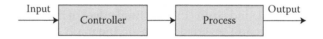

**Figure 1.5** Block diagram of an open-loop control system.

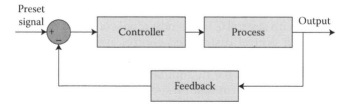

**Figure 1.6** Block diagram of a closed-loop control system.

vital parameters of a GT such as vibration, temperature, pressure, rotational speed, load, oil level and quality, and so on. Besides, it should be able to predict the future state of the system and to prevent unwanted shutdowns, as well as fatal breakdowns.

### 1.3.5.2   Fault detection and diagnosis

A GT model can be created in order to predict and detect faults in the system. Fault diagnosis acts as an important and effective tool when operators want to shift from preventive maintenance to predictive maintenance in order to reduce the maintenance cost [17]. It is concerned with monitoring a system to identify when a fault has occurred as well as to determine the type and location of the fault.

### 1.3.5.3   Sensor validation

GT models can be used for sensor validation purposes. Sensors are essential parts of any industrial equipment. Without reliable and accurate sensors, monitoring and control system of the equipment cannot work properly and may even face shutdown. Sensor validation is about detection, isolation, and reconstruction of a faulty sensor. It can improve reliability and availability of the system, and reduce maintenance costs. It enhances the reliability of the equipment and safety for the personnel. Sensor validation is also an effective tool to prevent unwarranted maintenance or shutdown. It has a considerable effect in increasing equipment's lifetime and assuring reliable performance. It can strengthen automation of the system by providing valid data for diagnostic and monitoring systems.

### 1.3.5.4   System identification

One of the main objectives of GT modeling is system identification. System identification infers a mathematical description that is a model of a dynamic system from a series of measurements of the system [18]. Despite significant research carried out in this field during the last decades, there is still a need for GT models with higher degree of accuracy and reliability for system identification purposes. This is due to the nonlinear and complex nature of GT dynamics.

### 1.3.5.5   Design and optimization of control system

GT models may be created to design or optimize GT control systems. It is obvious that any control system should be able to measure the output of the system using sensing devices and to take required corrective action if the value of measured data deviates from its desired corresponding value [19]. Control as a branch of engineering deals with the behavior of dynamical systems. The output performance of the equipment, which is under control, is measured by sensors. These measurements can be used to give feedback to the input actuators to make corrections toward desired performance. There are increasing demands for accurate dynamic models and controllers, in order to investigate the system response to disturbances and to improve existing control systems. Using new modeling methods can always be investigated as part of the optimization process.

## 1.3.6   GT model construction approaches

There are many sources regarding modeling and simulation of GTs. Various kinds of models have been built so far from different perspectives and for different purposes. Models of industrial systems can be classified into two main categories including black-box and white-box models.

### 1.3.6.1   White-box models

A white-box model is used when there is enough knowledge about the physics of the system. In this case, mathematical equations regarding dynamics of the system are utilized to make a model. This kind of model deals with dynamic equations of the system, which are usually coupled and nonlinear [20]. To simplify these equations in order to make a satisfactory model, making some assumptions based on ideal conditions and using different methods for linearization of the system is unavoidable. There are different software such as Simulink®, MATLAB®, and MATHEMATICA, which are extremely helpful, in this case.

### 1.3.6.2   Black-box models

A black-box model is used when no or little information is available about the physics of the system [20]. In this case, the aim is to disclose the relations between variables of the system using the obtained operational input and output data from performance of the system. Artificial neural network (ANN) is one of the most significant methods in black-box modeling. ANN is a fast-growing method, which has been used in different industries during recent years. The main idea for creating ANN, which is a subset of artificial intelligence, is to provide a simple model of the human brain in order to solve complex scientific and industrial problems in a variety of areas.

### 1.3.6.3   Gray-box models

In addition to white-box and black-box methods, the phrase gray-box may also be used when an empirical model is improved by utilizing a certain available level of insight about the system [18]. In this case, experiments can be combined with mathematical model building to improve model accuracy [20].

## 1.4  Problems and limitations

There is a rich source of research activities in the area of modeling, simulation, and control of GTs. However, in spite of all the efforts already done in this field, extensive attention still needs to be paid to this area in order to resolve problems encountered during the processes of design, manufacturing, operation, and maintenance of GTs. The following problems can be highlighted in the existing models and control systems of GT:

- Model and control methodologies, which are based on white-box approaches rely on thermodynamic and energy balance equations, which are coupled and have a high degree of nonlinearity. Therefore, considering assumptions and using linearization methods for simplification and solving these complex dynamics are unavoidable. Consequently, models and control systems that are built with such simplified and/or linearized equations are not accurate enough to capture system dynamics precisely. It, in turn, leads to unpredictable problems such as sudden shutdowns especially during operation of the GTs which are built based on or using such models. These facts demonstrate the necessity of using techniques and methodologies, which are independent of the system dynamics. Besides, the algorithm of proportional-integral-derivative (PID) controllers might be difficult to deal with in highly nonlinear and time-varying processes [21]. ANN has the capability to capture greatly complex dynamics of GTs quite independent of the physics of the system. Hence, the necessity of research in this area is obvious.
- GT components deteriorate gradually and lose their operability and efficiency with time. After a couple of years of service in industry, ideal thermodynamic relationships, and consequently the corresponding white-box models are subject to major changes and will not be valid anymore. Thus, prediction of behavior of old GTs is very difficult. On the other hand, replacement of old GTs by the new ones requires huge financial resources and in most cases is not economically viable. Fortunately, black-box models are very helpful in this case due to their independence and adaptability to the new conditions. Training and using an up-to-date ANN-based model for condition monitoring on the basis of new data sets of GT parameters can solve the problem.
- There has been little success in developing a dynamic model of GTs (and in particular for the start-up maneuver) by means of black-box approaches such as nonlinear autoregressive exogenous (NARX) models and to validate it against experimental data taken during the normal operation. Building the required models in this specific area

can be very effective in understanding and analyzing GT dynamics, and can also provide information about fault diagnostics. An NARX model, as a recurrent neural network (RNN), has the capability of capturing dynamics of complicated systems and can be employed for optimization of design and manufacturing of GTs. It can also be used to save time and money during the whole operation and maintenance period of GTs.

- There are little sources regarding comparison of physics-based models such as Simulink models with ANN-based models in terms of their deviations from the real systems. It is interesting and important to know how different methodologies work in this field and what their benefits and limitations are.
- Majority of available GT models have been built based on the steady-state operation of GTs. Further research needs to be carried out in the area of GT transient and start-up procedures.
- There is still a high demand for improving models and control systems, which are stable against changes in environmental conditions and system disturbances arising from faults or load fluctuations in the power network of IPGTs.
- Modeling, simulation, and control of GTs cover a broad range of activities. There are different types of GTs and a variety of modeling methods and control systems. Even in the field of ANN-based modeling and control, there are varieties of static and dynamic approaches and metrologies, which have not been investigated so far. Therefore, new developments in ANN-based structure need to be made for a variety of GTs types.

## 1.5   Objectives and scope

The main objective of this book is to provide and develop novel approaches and methodologies in modeling, simulation, and control of GTs for steady-state and start-up procedures by using ANNs. Steady-state and start-up operations of GTs are considered, physics-based and ANN-based models are built and compared on the basis of thermodynamic equations, energy balance relationships and mathematical analysis. Both simulated and experimental data are employed, and MATLAB tools including Simulink and Neural Network Tool-Boxes (NNTs) are used. The book has following objectives:

- Development of a novel ANN-based methodology for offline system identification of GTs based on combinations of various training functions, number of neurons, and transfer functions by using multilayer perceptron (MLP) structure. The model can be applied reliably for system identification of GTs and can predict output parameters of GTs based on the changes in inputs of the system with a high accuracy.

- Set-up and verification of a NARX model of start-up operation of an IPGT by using experimental time-series data sets. Comparisons are made between significant outputs of the model and the values of the corresponding measured data. The aim is to show that NARX models are capable of capturing system dynamics during start-up operation.
- Modeling and simulation of the transient behavior of an IPGT by employing Simulink and NARX approaches: The Simulink model is constructed based on the thermodynamic and energy balance equations in MATLAB environment. The measured time-series data sets are used to model operating characteristics and to make correlations between corrected parameters of the compressor and turbine components. The NARX model is set up based on the same data sets. Comparisons are made between significant outputs of the Simulink and NARX models with the values of the corresponding measured data. The objective is to demonstrate and compare capability of Simulink and NARX models for prediction of the transient behavior of GTs.
- Design of PID and ANN-based controllers for a single-shaft GT: Two different ANN-based control architectures including model predictive control (MPC), and feedback linearization control (NARMA-L2) are employed. The related control parameters are tuned according to the requirements of the design and comparisons are made among the performances of all controllers.

## 1.6   Summary

This chapter presented and discussed the motivations, necessity, and goal of modeling and control of GTs. A classification of GTs was provided, and main considerations in GT modeling were presented. The chapter briefly explained the most important criteria and considerations at the beginning of GTs modeling process including GT types and configurations, modeling methods, control system types and configurations, as well as modeling objectives and approaches. The chapter also defined the current problems in the area of modeling, simulation, and control of GTs. Finally, scope and main objectives of current and future activities in the area of GT modeling and simulation were briefly presented.

# chapter two

# White-box modeling, simulation, and control of GTs

> All the forces in the world are not so powerful as an idea whose time has come.
>
> **Victor Marie Hugo**
> *French poet, novelist, and dramatist, 1802–1885*

This chapter summarizes the most important studies carried out so far in the field of modeling, simulation, and control of GTs by using a white-box approach. White-box models of GTs can be categorized into low-power GT, IPGT, and aero GT models. In an IPGT, the mechanical power generated by the GT can be used by an alternator to produce electrical power. However, in an aero GT, the outgoing gaseous fluid can be utilized to generate thrust.

## 2.1   White-box modeling and simulation of GTs

### 2.1.1   White-box models of low-power GTs

A nonlinear state-space model of a low-power single-shaft GT for loop-shaping control purposes was developed by Ailer [22–24], and Ailer et al. [25–28]. The main idea of these studies was to improve dynamic response of the engine by implementation of a developed nonlinear controller. The model was developed and simulated in Simulink®-MATLAB® software, based on engineering principles, the GT dynamics, and constitutive algebraic equations. Model verification was performed by open-loop simulations against qualitative operation experience and engineering intuition. The researchers considered several assumptions during the modeling process in order to simplify the complicated nonlinear model and to obtain a low-order dynamic model. Although the assumptions made the model appropriate for control purposes, some important aspects of the GT dynamics were neglected during the simplification process.

Abdollahi and Vahedi [29] developed a dynamic model of single-shaft microturbine (SSMT) generation systems. They tried to present a general model that can be used in different operational ranges. The researchers emphasized on the functionality and accuracy of each MGT component and the complete model as well. They provided a dynamic model for each component of the microturbine including GT, DC bridge rectifier, permanent magnet generator, and power inverter. The models were implemented in Simulink-MATLAB. They showed that the models were suitable for dynamic analysis of microturbines under different conditions, and recommended that the model could also be useful to study the effect of microturbines on load sharing in power distribution network.

Aguiar et al. [30] investigated modeling and simulation of a natural gas-based micro turbine using MATLAB. The objective of the research was to present a technical and economical analysis of using MGTs for residential complex based on a daily simulation model and according to the environmental conditions. To evaluate the use of MGT for residential buildings, the researchers considered and analyzed two different configurations, based on the fact that the system was dimensioned to meet the thermal or the electrical demand. The results of the analysis could be useful for the investors who are interested to predict the cost of investment, operation, and maintenance of these turbines for power generation.

Ofualagba [31] presented modeling of a single-shaft MGT generation system suitable for power management in distributed generation applications. Detailed mathematical modeling of the control systems was investigated, and simulation of the developed MGT system was carried out by using Simulink/MATLAB. The developed model had the capability of matching with the power requirements of the load, within MGT's rating.

An approximate expression for part-load performance of a microturbine combined heat and power system heat recovery unit was identified by Rachtan and Malinowski [32]. They stated that the expression could greatly facilitate mathematical modeling, design, and operation of cogeneration plants based on MGTs and help in prediction of available thermal power. Malinowski and Lewandowska [33] explored an analytical model of an MGT for part-load operation. They calculated exergy destruction or loss for each MGT component. In thermodynamics, exergy is the theoretical desired output of a system during a process as it interacts to equilibrium [34]. They employed universal formulas with adjustable coefficients to overcome the problems caused by lack of information about the compressor and turbine performance maps, which are usually not disclosed by the manufacturers. The model was validated against and showed a good agreement with the experimental and manufacturer's data.

Hosseinalipour et al. [35] developed static and linear dynamic models of an MGT. They employed thermodynamic equations and maps of the MGT components to build the static model. Static and dynamic equations and a linearization methodology were used to set up the linear dynamic model. The models were validated, and a comparison was made between the results of the static model and those of the dynamic model for the steady-state condition.

## 2.1.2    White-box models of IPGTs

Rowen [36] presented a simplified mathematical model of a heavy-duty single-shaft GT. The objectives of his study were to investigate power system stability, to develop dispatching strategy, and to provide contingency planning for the system upsets [36]. Rowen tried to make a simplified model that could cover the full spectrum of GTs and appropriate turbine–generator characteristics. He discussed different issues regarding modeling including parallel and isolated operations, gas and liquid fuel systems, as well as isochronous and droop governors. The resulting model was very useful in studies related to power system dynamics. Although Rowen's model has been a base for many researchers to build up varieties of GT models using different approaches, it is limited to simple cycles and single-shaft GTs. He stated that engineering considerations and careful evaluation of the intended purpose are essential prior to the use of the model [36]. Rowen [37], in another effort, investigated a simplified mathematical model for the same GT with characteristics and features that affect the application of this kind of GT to mechanical drive services with a variable speed. The new features that were not included in his previous study [36] included calculation of exhaust flow, accommodation of variable ambient temperature, and modulating inlet guide vanes (IGVs). He intended to present a simple, but highly flexible and fairly accurate model. The characteristics of both fuel and control systems were incorporated into the model. Rowen's studies made it possible to simulate any heavy-duty single-shaft GT.

Najjar [38] investigated performance of GTs in single-shaft and twin-shaft operation modes using a model of a free power GT driving an electric dynamometer. GT operational data and their related curves for important parameters such as thermal efficiency, specific fuel consumption, and net output power were considered in order to estimate GT performance. The results showed that when the free PT engine was run in the single-shaft mode (especially with a low-speed ratio), power increased significantly (about 75% in the low-power region) at part loads. However, running the free PT engine in the two-shaft mode showed better torque characteristics at part load, which is really important for transport applications and traction systems [38].

Bettocchi et al. [39] and Binachi et al. [40] respectively explored dynamic models of single-shaft and multishaft GTs for power generation. An investigation into the use of exhaust gases of an open-cycle twin-shaft GT was performed by Mostafavi et al. [41]. They carried out a thermodynamic analysis and concluded that at low-temperature ratios, precooling could increase the efficiency and specific network of the cycle. Besides, depending on the cycle pressure ratio and the degree of precooling, the precooled cycle could operate at a higher compressor pressure ratio and temperature ratio without increasing the maximum cycle temperature.

A model for a twin-shaft GT was estimated by Hannett et al. [42]. They conducted a field testing program to obtain the required data for simulation of the model and assessment of the GT governor response to disturbances. During the process of model derivation, the model structure consisting of pertinent variables and parameters was determined. The researchers considered steady-state characteristics of the GT carefully in order to capture dynamic responses of important variables including rapid load changes and load rejections. To adjust the model parameters, an intelligent trial and error process was employed until reasonable matches were obtained between tests and simulations. This methodology was the only practical procedure for the model derivation because of the nonlinearity of the process and its controls. The researchers had to provide the required performance data for each GT component including the compressor, combustor, and turbines. Regardless of the complicated process of the model derivation, the resulting model could be useful for the studies of system dynamics.

A dynamic model for a twin-shaft GT was developed by Ricketts [43] based on a generic methodology and by using design and performance data. Because of the significant contribution of the effects of heat soak in the GT components to the dynamic characteristic of the GT they were included in the model. The model complexity was sufficient to predict transient performance and to facilitate designing an appropriate adaptive controller.

Crosa et al. [44] explored a nonlinear physical model to predict the off-design and steady-state dynamic behavior of a heavy-duty single-shaft GT using Simulink-MATLAB. They used dynamic equations of mass, momentum, and energy balances to model the system. The air bleed cooling effect, the mass storage among the stages, and the air bleed transformations from the compressor down to the turbine were taken into account for the model building process. Performance of the resulting model was quite satisfactory for the prediction of thermodynamic variables.

Nagpal et al. [45] presented their field experiences in testing and validation of turbine dynamic models and the associated governors for IPGTs. Based on the field measurements, they showed that gas turbine governor model (GAST), which is a widely used model to represent the dynamics of GT governor systems, has two main deficiencies. First, the model could not predict GT operation accurately at high levels of loads. Second, the accurate adjustment of the model parameters, according to the oscillations around the final setting frequency, may not be attained.

Kaikko et al. [46] presented a steady-state nonlinear model of a twin-shaft industrial GT and its application to online condition monitoring and diagnostic system. They utilized condition parameters to evaluate the engine condition and the impact of performance deviations on the costs. Using the condition parameters, the performance was predicted by the reference operating conditions for the engine with the current health status. Evaluation of the GT performance parameters in reference, actual, and expected and corrected states enabled the researchers to properly identify the deviations and their root causes. They also concluded that the applied computational method in their study could be adapted to other modeling, condition monitoring, and diagnosis of GTs. The methodology employed by the researchers had some advantages compared with the commonly applied component matching procedures. It facilitated the selection of the modeling parameters as well as application of the models for providing and controlling the results.

Al-Hamdan et al. [47] discussed modeling and simulation of a single-shaft GT engine for power generation based on the dynamic structure and performance of its individual components. They used basic thermodynamic equations of a single-shaft GT to model the system. The researchers developed a computer program for the engine simulation, which could be used as a useful tool to investigate GT performance at off-design conditions and to design an appropriate, efficient control system for specific applications. Figure 2.1 shows variations of temperatures in different sections of the modeled GT versus net power output. $T_{02}$, $T_{03}$, and $T_{04}$ are output temperatures of compressor, combustor, and turbine respectively [47].

A simplified desktop performance model of a typical heavy-duty single-shaft GT in power generation systems was developed by Zhu and Frey [48]. They built a model which could be accurate and robust to variations under different operational conditions. The researchers investigated a methodology for assessment and rapid analysis of the system alternations.

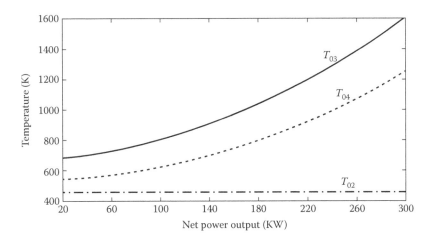

*Figure 2.1* Variations of temperatures versus net power output. (From Q. Z. Al-Hamdan and M. S. Y. Ebaid, *ASME Journal of Engineering for Gas Turbines and Power,* vol. 128, no. 2, pp. 302–311, 2006.)

The methodology could be implemented in a desktop computing environment. They applied sensitivity analysis to assess the model for a variety of fuels in terms of composition, moisture, and carbon contents. The model could also be used to evaluate $CO_2$ emissions.

Camporeale et al. [49] investigated an aerothermal model for two different power plant GTs with a relatively high level of accuracy. They presented a novel methodology for developing a high-fidelity real-time code in Simulink-MATLAB using an object-oriented approach for GT simulation. The technique was based on a nonlinear representation of GT components. The researchers composed and solved a set of ordinary differential equations and nonlinear algebraic equations to present the mathematical model of the GTs. The flexibility of the code allowed it to be easily adapted to any configuration of power plants. Figure 2.2 shows how the real-time simulation software interacted with hardware control devices of the GTs [49].

A GT fully featured simulator was developed and implemented by Klang and Lindholm [50]. They discussed the simulator set-up both technically and economically, and chose a robust hardware solution based on the basic requirements. The simulator could be useful for testing the GT control system, trying out new concepts, and training operators.

Development of a dynamic model of a single-shaft GT for a combined-cycle power plant (CCPP) was explored by Mantzaris and Vournas [51]. They used Simulink-MATLAB to investigate the stability of the turbine and its control system against overheat, as well as changes in frequency

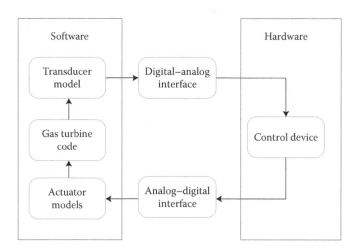

*Figure 2.2* **(See color insert.)** The diagram for how real-time simulation software interacted with hardware control devices. (From S. M. Camporeale, B. Fortunato, and M. Mastrovito, *ASME Journal of Engineering for Gas Turbines and Power*, vol. 128, no. 3, pp. 506–517, 2006.)

and load. The results showed that the existence of speed, frequency, and air control loops were necessary for the plant stability against disturbances. To make the model response faster, the researchers ignored some blocks with small time constants in the model for reducing the order of the model and simplifying the calculations. To allow stable and reliable operation of the plant, it was also suggested that the airflow gate opening limits be expanded during the full-load operation.

Yee et al. [52,53] carried out a comparative analysis and overview of different existing models of power plant GTs. They identified, presented, and discussed various kinds of GT models in terms of their application, accuracy, and complexity. It was concluded from the research that despite their complexity, physical models are the most accurate and suitable for detailed study of the GT dynamics. However, it was stated that physical models are not appropriate for the use in large power system studies. It was also indicated that for a more detailed analysis of power systems and their governors' behavior, the frequency-dependent model was the best choice. It was particularly useful in the case of having systems with large frequency variations. The study also demonstrated that the frequency and ambient temperature could significantly affect GT operation under certain operating conditions. Unfortunately, the study did not cover black-box models of GTs [52]. A similar study was carried out by Shalan et al. [54] for the GTs in CCPPs. They performed a complementary and comparative analysis of different GT models response in terms of their applications and accuracy. In another effort, Liang et al. [55] carried out a study on performance simulative models of GTs. The objective was to improve accuracy of nonlinear simulation models of GTs. They explored the influence of variations of oil to gas ratio, specific heat, and PT outlet pressure on the engine, in order to establish a nonlinear dynamic simulative model of twin-shaft GTs. The simulation results showed that the proposed simulative model was more accurate to reflect the engine dynamics compared to the previous ones. Hosseini et al. [56] employed a systematic methodology to build a multiple model structure of a prototype IPGT under normal operation. According to the methodology, linear and nonlinear modes were decomposed and treated separately. They concluded that the algorithm could be employed for the identification of single-shaft IPGTs.

A zero-dimensional simulation model for design and off-design performance of a twin-shaft GT was developed by Lazzaretto and Toffolo [57]. The aim of the study was to manage correctly, the operation of power plant GTs and their reactions to the variations of load and ambient temperature. The researchers determined the values of thermodynamic quantities and the overall performances of the GT plant. To predict nitrogen oxide ($NO_X$) and carbon monoxide ($CO$) pollution, available semiempirical correlations for pollutant emissions were adapted by tuning their coefficients on the experimental data. The researchers concluded that the applied methodology could be employed to manage the economical and environmental aspects of the plant operation.

Razali [58] developed an analytical model of a GT to simulate the actual trend of the GT performance and to predict its degradation. The actual composition of the working gases and variation of the specific heat with temperature are taken into account for simulating the model. The values of three output parameters from the resulting model including GT exhaust temperature, GG exit temperature, and the actual load, were compared with the corresponding actual outputs, and the deviations were measured as indicators of the GT degradation.

A modified methodology was presented by Khosravi-el-hossani and Dorosti [59] to determine the exhaust energy in the new edition of ASME PTC 22, which is about flow rate of the flue gas. The method was based on exhaust gas constituent analysis and combustion calculations. It was shown that the method could enhance the precision of ASME PTC 22 by more than 1%. The GT performance test was also improved based on the obtained operational data. They stated that the proposed methodology could be an appropriate alternative for GT standard performance test and could be employed to evaluate GT performance without measurement of input fuel components, which could reduce the cost of measuring and data gathering.

In a couple of different efforts, Ibrahim and Rahman [60,61], and Rahman et al. [62,63] developed computational models of a power plant GT in MATLAB environment. They investigated the effect of operational conditions (compressor ratio, air to fuel ratio, turbine inlet and exhaust temperatures, and efficiency of compressor and turbine) on the power plant performance (compressor work, heat rate, thermal efficiency, and specific fuel consumption). It was observed that the output power and thermal efficiency decreased linearly with increase of both ambient temperature and air to fuel ratio. They also concluded that the peak power, efficiency, and specific fuel consumption occurred at a higher compressor ratio with low ambient temperature.

The parameters of a single-shaft HDGT were estimated using its operational data based on Rowen's model [36] by Tavakoli et al. [64]. They applied simple physical laws and thermodynamic assumptions in order to derive the GT parameters using operational data. They suggested that the study could be useful for educational purposes especially for the students and trainers who were interested in GT dynamics. Figure 2.3 shows the block diagram of Rowen's model including fuel and control systems, employed by the researchers [36,64].

Simple models of the systems for a power plant simulator were developed by Roldan-Villasana et al. [65] based on the mass, momentum, and energy principles. The modeled systems were classified into seven main groups including water, steam, turbine, electric generator, auxiliaries, GT,

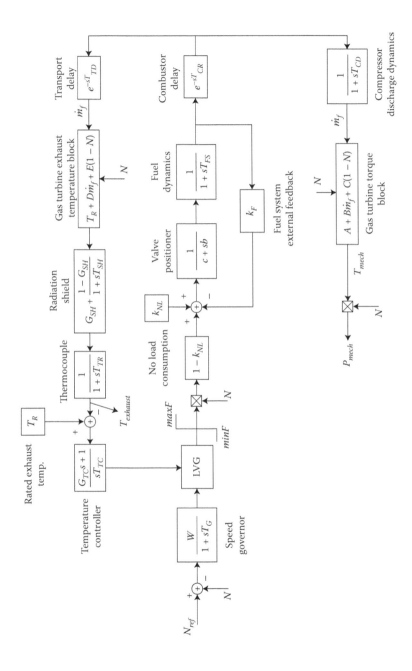

*Figure 2.3* Rowen's model for heavy-duty GTs dynamics. (From W. I. Rowen, *ASME Journal of Engineering for Power*, vol. 105, no. 4, pp. 865–869, 1983; M. R. Bank Tavakoli, B. Vahidi, and W. Gawlik, *IEEE Transactions on Power Systems*, vol. 24, no. 3, pp. 1366–1374, 2009.)

and minimized auxiliaries. They concluded that the simulator could be very useful for training of operators.

Yadav et al. [66] applied graph networks approach to analyze and model a single-shaft open-cycle GT. They used graph theory and algorithms to identify pressure and temperature drops, work transfer rates, rate of heat, and other system properties. Because of the similarities between the results from this approach and the results from conventional methods, it was suggested that the new technique could be used for optimization of GT process parameters.

Weber [67] investigated modeling of a modern power plant GT engine performance at part-load. At the first step, thermodynamic characteristics at full-load for the engine was employed in commercial software including MATLAB and Python, and then part-load performance thermodynamic characteristics were derived by using a computer programming code that was entirely flexible while being computationally efficient.

Shalan et al. [68], in another effort, employed a simple methodology to estimate parameters of a Rowen's model [36,37] for heavy-duty single-shaft GTs. The parameters of the model were derived using the performance and operational data. A variety of simulated tests was performed in Simulink-MATLAB environment, and the results were compared with and verified against the results of previous studies. The researchers stated that the proposed methodology could be applied to any size of GTs.

Liu and Su [69] developed a nonlinear model of an IPGT for faults diagnosis purposes. The GT was a part of a combined cycle generation unit. The objective of the research was to build a health monitoring-based thermodynamic model to explain quantitatively the degradation phenomenon in the gas path components of GTs. By using the component level nonlinear model, errors caused by linearization could be avoided. The dynamic model could evaluate steady-state behavior of the GT for off-design performance. The results showed that faults could be detected and isolated by using a model-based gas path analysis.

The effect of ambient temperature on performance of three-shaft GTs, under different control strategies, was investigated by Gao and Huang [70]. They showed that the ambient temperature greatly affected the GTs performance. They demonstrated that GT specific power and thermal efficiency, particularly when the GT worked in off-design conditions, decreased as the ambient temperature increased. They also concluded that the effect of variations in ambient temperature on three-shaft GTs was different under different work conditions, and suitable relevant factors should be considered choosing the appropriate control strategy.

Memon et al. [34] investigated a model of simple and regenerative cycle GT power plants. The objective of the research was to optimize the cycles for maximization of "net power output, energy, and exergy efficiencies" and minimization of "$CO_2$ emissions and costs of the cycles."

To estimate the response variables with a high degree of accuracy, the model equations were developed through regression analysis. The results showed that the regenerative cycle had smaller exergy destruction rate and thus more efficiency for a given operating condition compared with the corresponding values in the simple cycle.

To investigate the potential possibilities for improvement of part-load efficiency of GTs operating under variable speed, dynamic GT model for both single-shaft and twin-shaft engines were explored by Thirunavukarasu [71]. For this purpose, the mathematical models of the individual GT components were developed on the basis of thermodynamic laws, and the resulting model was validated for the design, off-design, and transient cases by using the available data. Besides, to explore the dynamic potential interaction between the GT operation and the electrical and thermal systems, the engine model was integrated with power generation, distribution, and thermal systems. Moreover, a variable speed parametric study is performed by using the developed GT model. The results showed that the efficiency increased as the load decreased and that the improvement of efficiency for single-shaft engines was larger compared with twin-shaft engines.

Shaw et al. [72] presented a GT-based model of a CCPP by using actual existing data. They explored effects of variations of ambient temperature and used operational data to validate the model. The results showed that changes in ambient temperature heavily affected the performance (particularly the output power) of the GT part of the GT cycle, but its effects on the performance of the steam cycle was almost negligible. They concluded that CCPP operation is more stable than a stand-alone GT in hot weather in summer.

Wiese et al. [73] developed a physics-based dynamic model of a GT and validated it against transient test data. It was concluded that the overall system dynamics could be captured well, and the dynamic model could be used in a model-based GT controller.

Al-Sood et al. [74] explored an irreversible GT Brayton cycle by developing a general mathematical model. The cycle incorporated compressor, GT, intercooler, reheater, and regenerator. They proposed a general mathematical formula which showed the effect of each of the operating parameters on the GT performance. They stated that the formula could be applicable under any operational conditions of the cycle regardless of values of the other parameters.

### 2.1.3   White-box models of aero GTs

Kim et al. [75] developed a model for a single-shaft turbojet engine using Simulink-MATLAB. The transient behavior and changes of different engine parameters were predicted by the model based on variations of the fuel flow rate. The researchers considered different flight conditions in their simulation such as fuel cut-off. Comparison of the simulation output with another dynamic code for GTs showed satisfactory results.

Evans et al. [76] examined a linear identification of fuel flow rate to shaft speed dynamics of a twin-shaft GT, which was a typical military Rolls-Royce Spey engine. They studied direct estimation of s-domain models in the frequency domain and showed that high-quality models of GTs could be achieved using frequency-domain techniques. They discussed that the technique might be used to model industrial systems, wherever a physical interpretation of the model is needed. In another effort, Evans et al. [77] presented the linear multivariable model of a twin-shaft aero GT, a typical Rolls-Royce Spey military turbofan, using a frequency-domain identification technique. The technique was employed to estimate s-domain multivariable models directly from test data. The researchers examined the dynamic relationship between fuel flow rate and rotational speed in the form of single-input and multiple-output (SIMO). The main advantage of the model was its capability to be directly compared with the linearized thermodynamic models. The research showed that a second-order model could present the most suitable model and the best estimation of the engine. The researchers suggested that the techniques investigated in their study could be used to verify the linearized thermodynamic models of GTs. Figure 2.4 shows the Rolls-Royce Spey engine modeled by Evans et al. [77].

Arkov et al. [78] employed four different system identification approaches to model a typical aircraft GT using the data obtained from a twin-shaft Rolls-Royce Spey engine. The motivation behind their research was to minimize the cost and to improve the efficiency of GT dynamical testing techniques. The four employed techniques by the researchers included "multi-sine and frequency-domain techniques for both linear and nonlinear models," "ambient noise excitation," "extended least-squares algorithms for finding time-varying linear models," and "multi-objective genetic programming for the selection of nonlinear model structures" [78]. A description of each technique and the relative merits of the approaches were also discussed in the study. In another effort, Arkov et al. [79] discussed a life cycle support for dynamic modeling of aero-engine GTs. They investigated different mathematical models and their applications at life cycle stages of the engine controllers. They developed a unified information technology and a unified information space for creating and using GT mathematical models at the life cycle stages. Standard methodologies for system modeling and appropriate software were employed for implementation of this new concept, and consequently performance enhancement of the control system.

Riegler et al. [80] explored modeling of compressor behavior in GT performance calculations by using a methodology for extrapolating the compressor maps. The research covered the extreme part-load regime of compressor operation. Using corrected torque instead of efficiency in the

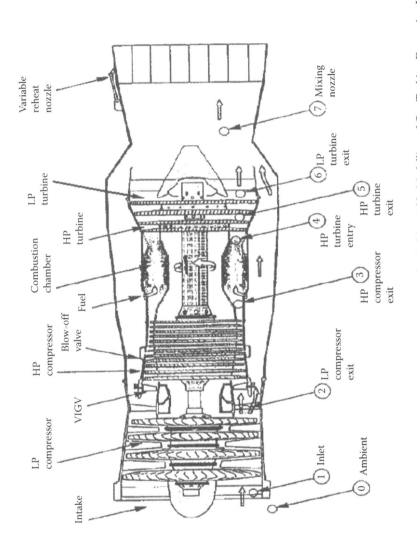

*Figure 2.4* A typical Rolls-Royce Spey engine. (From C. Evans et al., *Multivariable Modelling of Gas Turbine Dynamics*, University of Glamorgan & Vrije Universiteit Brussel, 2001.)

maps could facilitate calculation of GT behavior for the operating conditions. The researchers stated that the methodology employed in their study could also be used for typical turbomachinery relationships in turbines.

Behbahani et al. [81] employed Simulink-MATLAB to develop a nonlinear dynamic model of a two-shaft turbine engine for diagnostics and prognostics purposes. The model has the capability to be adapted successfully to various turbine engines. They also designed a controller to control the rotor speed. A survey of prognostic techniques for turbine engines was also carried out in the research.

A novel dimensionless modeling approach for prediction of performance of aero GT engine parameters was proposed by Pourfarzaneh et al. [82]. They set up a systematic series of experiments on the engine to obtain functional parameters of the GT components. The results showed an excellent agreement between theoretical and experimental values. A new flexible analytical methodology for linearization of an aero GT engine model was investigated and developed by Chung et al. [83]. Abbasfard [84] explored a modified linear multiple models for a single-shaft aero GT engine by using a novel symbolic computation-based methodology for linearization of the system. The simulation results showed that the proposed method had the capability of resolving fault detection and isolation (FDI) problems of the engine.

Lu et al. [85] proposed a model of an aero GT engine for sensor fault diagnostics purposes. The architecture of the model was composed of two nonlinear engine models including real-time adaptive performance and on-board baseline models. They also presented a novel approach to sensor fault threshold based on the model errors and noise level. The researchers concluded that the proposed approach was easy to design and tune with long-term engine health degradation. In another effort, Lu et al. [86] employed a model-based approach for health parameters estimation of an aero GT engine and stated that the suggested methodology was efficient.

## 2.2   White-box approach in control system design

Modeling and simulation of GTs play a significant role in control areas. GT models can improve GT control systems and reduce costs associated with the implementation of controllers on real systems. Different control strategies and a variety of controllers can be employed and tested on GT models before implementation on real systems. This section explains the main white-box based research activities in control of GTs.

Ricketts [43] showed that the dynamic model developed for a twin-shaft GT by using a generic methodology and performance data sets could represent an ideal application to adaptive control schemes. Ailer [87] and Ailer et al. [88–91] carried out different research to design and develop control systems for a low-power industrial GT based on the results they already achieved in nonlinear modeling of the GT. They used nonlinear modeling methodology based on thermodynamical equations to model the system. They linearized the model to be able to design different kinds of controllers. Agüero et al. [92] applied modifications in a heavy-duty power plant GT control system. One of the modifications limited speed deviations to the governor, which in turn limited power deviation over dispatch set point. Another modification could prevent nondesired unloading of the turbine. The researchers investigated turbine dynamic behavior before and after the modifications were made.

Centeno et al. [93] reviewed typical GT dynamic models for power system stability studies. They discussed main control loops including temperature and acceleration control loops and their applications and implementations. They also explained different issues, which should be considered for modeling of temperature and acceleration control loops. The performance of the control loops was simulated against changes in GT load. Figure 2.5 shows the block diagram of the basic temperature control loop for the GT model [93].

Ashikaga et al. [94] carried out a study to apply nonlinear control to GTs. They reported two applications of nonlinear control. The first one was the starting control using the fuzzy control, and the other was the application of the optimizing method to variable stator vane (VSV) control. The objective was to increase thermal efficiency and to decrease nitrogen oxide emission. However, the algorithms for solving optimization problems were complicated, time-consuming, and too large to be easily installed in computers. Zaiet et al. [95] proposed modeling and nonlinear control of a GT based on the previous studies. They stated that their methodology could provide more flexibility in design of strategies,

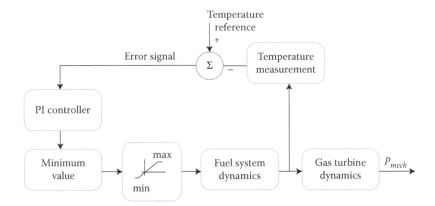

*Figure 2.5* Block diagram of the basic temperature control loop for a GT model. (From P. Centeno et al., *Review of Gas Turbine Models for Power System Stability Studies,* Madrid: Universidad Pontificia Comillas and Endesa Generación, 2002.)

controlling the speed and surge simultaneously, and accelerating the compressor without stalling problems.

Lichtsinder and Levy [96] developed a simple and fast linear model for real-time transient performance of a jet engine control. They formulated control system specifications to specify the maximal variance of the fuel flow command during transient maneuvers. They defined and employed a novel generalized function definition and discussed the application of this technique for the development of the model. The resulting model showed a high accuracy for variance of the fuel flow rate. Comparison of the simulation results to the conventional models showed that the new model could successfully be used for large input variances.

Pongraz et al. [97] used an input–output linearization method to design an adaptive reference tracking controller for a low-power GT model. They discussed a third-order nonlinear state space model for a real low-power single-shaft GT based on the dynamic equations of the system. In their model, fuel mass flow rate and rotational speed were considered as input and output, respectively. A linear adaptive controller with load torque estimation was also designed for the linearized model. According to the results of the simulation, the required performance criteria were fulfilled by the controlled plant. The sufficient robustness of the system against the model parameter uncertainties and environmental disturbances were also investigated and approved.

Tong and Yu [98] presented a dynamic model of a microturbine and its nonlinear PID controller. Their research objective was to improve the stability of the microturbine system. The simulation results demonstrated that although the nonlinear PID controller has better adaptability and robustness than the classical PID, microturbine system could not operate smoothly at all operating modes. Therefore, they suggested that for better performance, it was necessary to use both auto-disturbance rejection control and nonlinear robust, coordinated control methodologies.

Najimi and Ramezani [99] designed a robust controller for an identified model of a power plant GT. The applied model was built based on Rowen's model [37] and by using real data for tuning the GT parameters. The controller design objectives were to adjust the GT rotational speed and exhaust gas temperature simultaneously by controlling fuel signals and compressor IGV position. Simulation results showed that the proposed controller for the nonlinear model of the system fairly fulfilled the predefined objectives as it could maintain turbine speed and exhaust gas temperature within the desired interval, under load disturbances and nonlinear uncertainties. Compared with MPC and PID controllers, the robust controller decreased maximum amplitude of the speed deviations remarkably.

Kolmanovsky et al. [100] developed a robust control system for aero GTs. The purpose of the study was to preserve stability and tracking performance of the engine under uncertainties such as surge margins and large inlet distortions. To develop stable control architectures for GT engines, Pakmehr et al. [101] investigated a nonlinear physics-based dynamic model of a twin-shaft aero engine. A stable gain scheduled controller was presented, and a stability proof was explored for the closed-loop control system. Besides, gain scheduled model reference adaptive control for multiple-input multiple-output (MIMO) nonlinear plants with constraints on the control inputs was studied.

There have also been remarkable academic research efforts, trying to develop a theoretical background of MPCs. An overview of industrial MPC technology was presented by Qin et al. [102,103]. Richalet [104,105] discussed industrial applications of MPC. A pedagogical overview of some of the most important developments in MPC theory and their implications for the future of MPC theory and practice was discussed by Nikolaou [106]. Rawlings provided a review of MPCs for tutorial purposes [107].

## 2.3   Final statement

As it can be seen, the outcome of the research in the field of white-box modeling, simulation, and control of GTs has been very effective in GT performance evaluation and optimization before final design and manufacturing processes. However, there is still a great need for further system optimization. To approach an optimal model as closely as possible, researchers need to unfold the unknowns of complicated nonlinear dynamic behavior of these systems in order to minimize undesirable events such as unpredictable shutdowns, overheating, and overspeed during the GT operation. Therefore, further research and development activities need to be carried out in this field.

In addition, majority of both white-box and black-box models in the literature have been built based on the steady-state operation of GTs, when GTs have already passed the start-up procedure and run in a stable mode. Unfortunately, the literature lacks enough investigation about modeling and simulation of GT transient behavior and start-up operation, especially for IPGTs. Among the limited number of studies covering this topic, one can refer to the works by Agrawal and Yunis [108], Balakrishnan and Santhakumar [109], Peretto and Spina [110], Henricks [111], Beyene and Fredlund [112], Kim et al. [113–115], Shin et al. [116], Davison and Birk [117], Huang and Zheng [118], Xunkai and Yinghong [119], Sanaye and Rezazadeh [120], Kocer [121], Corbett et al. [122], Alobaid et al. [123], Zhang [124], Rezvani et al. [125], Daneshvar et al. [126], Rahnama et al. [127], Refan et al. [128], and Sarkar et al. [129]. There are also some white-box and black-box methodologies regarding simulation of transient behavior of individual main components of GTs such as compressors, which can be effectively applied to GTs. For instance, one can refer to NN techniques employed by Venturini [130,131], to explore transient behavior of compressors. Similar efforts were carried out by Venturini [132], and Morini et al. [133,134] by using white-box methods. Therefore, because of the importance and sensitivity of these procedures in GT service life, further research is needed to be carried out in this area.

Since it is desirable to design GTs with high performance, high reliability, and cost effectiveness, an extensive effort still needs to be devoted toward understanding their complex natural dynamics and coupled parameters. For instance, system disturbances arising from faults or load fluctuations in the power network of power plant GTs may drive GTs to instability. Exploring reaction of GTs to the system disturbances and changes in environmental conditions is still a challenging issue. Therefore, there is an increasing demand for accurate dynamic models, to investigate the system response to disturbances and to improve existing control systems. The investigated models and control systems which were built based on simplified and/or linearized equations are not accurate enough to capture system dynamics precisely. Therefore, a thorough analysis needs to be carried out regarding the problems that linearization may cause for modeling and control of GTs [22,135]. Application of ANN as a fast and reliable method to stabilize the system against disturbances can be investigated further. In this case, dynamic behavior of the system can be predicted and controlled using novel control methods in the presence of a number of uncertainties, such as environmental conditions and load changes.

It is not possible to provide detailed discussion on the entire research directory in this book. Although aero GTs and stationary GTs have the same basic structures, they have many differences in details. Because of high demand of electricity market for optimization of power plant GTs and the need for extensive research in modeling and control of these types of GTs, special attention needs to be paid to this area.

## 2.4  Summary

This chapter presented a comprehensive overview of the research activities in the field of white-box modeling, simulation, and control of GTs. It discussed most relevant scientific sources and significant research activities in this area for different kinds of GTs. Main white-box models and their applications to control systems were investigated for low-power, aero, and power plant GTs. It was shown that despite significant studies in this area, further research needs to be carried out to resolve unpredictable challenges that arise in the manufacturing processes or in the operation of industrial plants. These challenges may be found in a variety of areas such as design, commissioning, condition monitoring, fault diagnosis, troubleshooting, maintenance, sensor validation, control, and so on.

# chapter three

# Black-box modeling, simulation, and control of GTs

> All intelligent thoughts have already been thought;
> what is necessary is only to try to think them again.
>
> **Johann Wolfgang von Goethe**
> *German writer and politician, 1749–1832*

This chapter summarizes the most important studies carried out so far in the field of modeling, simulation, and control of GTs by using black-box approach. As in white-box models, black-box models can be categorized into low-power GT, IPGT, and aero GT models.

## 3.1 Black-box modeling and simulation of GTs

### 3.1.1 Black-box models of low-power GTs

A NARX model was employed by Jurado [136] to model a power plant MGT and its related distribution system dynamics. However, the nonlinear terms in the model were restricted to the second order. The modeling objective was to investigate the impacts of this kind of GT on the transient and long-term stability of the future distribution systems. The resulting model was capable of modeling both low and high amplitude dynamics of MGTs. The quality of the model was examined by cross-validation. The model was tested under different operational conditions and electrical disturbances. The results showed that NARX methodology could be applied successfully to model MGT dynamics in nonisolated mode [136].

Application of ANN and adaptive network-based fuzzy inference system (ANFIS) to MGTs was presented by Bartolini et al. [137]. They used ANN and ANFIS to explore unavailable experimental data in order to complete the MGT performance diagrams. They also analyzed and predicted emissions of pollutants in the exhausts and investigated the effects of changes of ambient conditions (temperature, pressure, and humidity) and load on MGT's output power. The results indicated that ANN could effectively assess both MGT performance and emissions. It was also shown that ambient temperature variations had more effect on the output power than humidity and pressure. Besides, MGTs were less influenced by ambient conditions than load.

Nikpey et al. [138] developed an ANN-based model for monitoring of combined heat and power MGTs by using the data collected from a modified MGT on a test rig. A systematic four-step sensitivity analysis was performed to investigate the relevance of the input and output parameters, as well as influence of input parameters on the prediction accuracy of each output. The results demonstrated that the compressor inlet measurements had very significant impacts on improvement of the prediction accuracy of the model, and could also act as representatives of ambient measurements, so that ambient measurements could be excluded from the inputs. The results of sensitivity analysis also showed that compressor inlet temperature, compressor inlet pressure, and power were the most significant input parameters. The final result indicated that the ANN model could predict the normal performance of the MGT with high reliability and accuracy.

## 3.1.2   Black-box models of IPGTs

Lazzaretto and Toffolo [139] investigated a zero-dimensional design and off-design modeling of a single-shaft GT using ANN. They used analytical method and feedforward neural network (FFNN) as two different approaches to predict GT performance. Appropriate scaling techniques were employed to construct new maps for the GT using available generalized maps of the compressor and turbine. The new maps were validated using the experimental data obtained from real plants. Off-design performance of the GT was obtained using modifications of the compressor map according to variable inlet guide vane closure. A commercial simulator was employed to solve the set of equations of the developed analytical model. Different sets of independent variables that could be selected according to the available data allowed a high flexibility in the choice of the adjustment criteria. However, the effects of internal parameter variations on GT performance were not considered in the analytical approach. The results from the simulator were used for training the FFNN. The resulting ANN model showed excellent prediction accuracy with just about 1% error. The researchers emphasized the reliability of the ANN model in making accurate correlations between important thermodynamic parameters of complex thermal systems.

Ogaji et al. [140] applied ANN for multisensor fault diagnosis of a stationary twin-shaft GT using Neural Network Tool-Box in MATLAB®. The GT performance was thermodynamically similar to the Rolls-Royce Avon engine. The required data for training the networks were derived from a nonlinear aero-thermodynamic model of the engine's behavior. The researchers presented three different ANN architectures. The first ANN

was used to partition engine measurements into faults and no-faults categories. The second network was employed to classify the faults into either
a sensor or a component fault. The third ANN was applied to isolate any
single or dual faulty sensors and then to quantify the magnitude of each
fault, via the difference between the network's inputs and outputs. The
results indicated that ANN could be used as a high-speed, powerful tool
for real-time control problems [140].

Arriagada et al. [141] applied ANN for fault diagnosis of a single-shaft
industrial GT. They obtained data sets from 10 faulty and 1 healthy engine
conditions. The data sets were employed to train a feedforward MLP neural network. The trained network was able to make a diagnosis of the
GTs condition. The results showed that ANN could identify the faults and
generate warnings at early stages with high reliability. Figure 3.1 shows a
schematic drawing of the ANN and the interpretation of the outputs [141].

As it can be seen from the figure, the inputs correspond to the 14
measured parameters in the real engines, as well as the ones controlled
by the operators and the control system. The parameters include ambient temperature, IGV angle, mass flow rate, fuel flow rate, load, pressure,
temperature, and so on. The desired outputs from the ANN are unique
combinations of 28 binary numbers arranged in a graphical display. The
training process of the ANN stopped when it showed the best performance based on a selected number of hidden neurons and weights. The
ANN could be named 14-H-28 according to its structure [141].

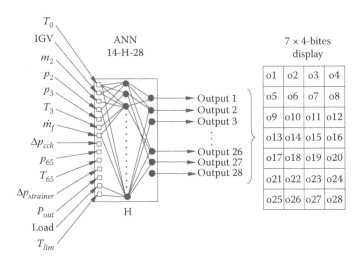

*Figure 3.1* A schematic drawing of the ANN model and the interpretation of the
outputs in a graphical display. (From J. Arriagada et al., Fault diagnosis system for
an industrial gas turbine by means of neural networks, in *International Gas Turbine
Congress*, Tokyo, November 2–7, 2003, 6 pp.)

Basso et al. [142] applied an NARX model to identify dynamics of a small heavy-duty IPGT. Their objective was to make an accurate reduced-order nonlinear model using black-box identification techniques. They considered two operational modes for the GT; when it was isolated from the power network as a stand-alone unit and when it was connected to the power grid. The parameter estimation of the NARX model was performed iteratively using Gram–Schmidt procedure. Both forward and step-wise regressions were investigated, and many indices were evaluated and compared to perform subset selection in the functional basis set and to determine the structure of the nonlinear model. A variety of input signals was chosen for system identification and validation purposes.

Kaiadi [143] developed an ANN-based modeling of a heat and power plant for monitoring and performance analysis purposes by employing commercial software for training cross-validation and testing processes. The model-building process was divided into two submodules. Two different ANN models were developed separately and then linked to each other. To make the data set as reliable as possible, data preprocessing was carried out before the training stage.

Bettocchi et al. [144] investigated an ANN model of a single-shaft GT as an alternative to physical models. They tried to explore the most appropriate neural network (NN) model in terms of computational time, accuracy, and robustness. The researchers considered a network with 15 inputs and 6 outputs. The required data sets for training of the network were obtained from a cycle program, previously calibrated on the GT engine. The obtained data covered the entire operational range of the GT and the researchers considered different health states. They concluded that a feed-forward MLP with a single hidden layer (including 60 neurons) trained with at least 2000 training patterns was the most appropriate network. They observed that ANN could be very useful for the real-time simulation of GTs especially when there was not enough information about the system dynamics. In a similar study, Bettocchi et al. [145] developed a MIMO NN approach for diagnosis of single-shaft GT engines. In another research, Spina and Venturini [146] used field data sets and applied ANN to train operational data through different patterns in order to model and simulate a single-shaft GT and its diagnostic system with a low computational and time effort.

Simani and Patton [147] used a model-based approach to detect and isolate faults on a single-shaft industrial GT prototype. They suggested exploiting an identified linear model in order to avoid nonlinear complexity of the system. For this purpose, black-box modeling and output estimation approaches were applied due to their particular advantages in terms of algorithmic simplicity and performance achievements. The suggested fault diagnosis strategy was especially useful when robust solutions were required for minimizing the effects of modeling errors and noise while maximizing fault sensitivity. To verify the robustness of the obtained solution, the proposed FDI approach was applied to the simulated data from the GT in the presence of measurement and modeling errors. Yoru et al. [148] examined application of the ANN method to exergetic analysis of GTs which supplied both heat and power in a cogeneration system of a factory. They compared the results of the ANN method with exergy values from the exergetic analysis and showed that much closer exergetic results could be attained by using the ANN method.

Fast et al. [149] applied simulation data and ANN technique to examine condition-based maintenance of GTs. In another effort, Fast et al. [150] used real data obtained from an industrial single-shaft GT working under full load to develop a simple ANN model of the system with a very high prediction accuracy. A combination of ANN method and cumulative sum (CUSMUS) technique was utilized by Fast et al. [151] for condition monitoring and detection of anomalies in GT performance. Application of ANN to diagnosis and condition monitoring of a combined heat and power plant was discussed by Fast et al. [152]. Fast applied different ANN approaches for GT condition monitoring and sensor validation and diagnosis [153]. To minimize the need for calibration of sensors and to decrease the percentage of shutdowns due to sensor failure, an ANN-based methodology was developed for sensor validation in GTs by Palmé et al. [154]. Nozari et al. [155] employed MLP models for an IPGT, based on a nonlinear dynamic system identification approach to detect and isolate the GT faults. The proposed method for fault detection and isolation was tested and validated on a single-shaft IPGT. Besides, to show the benefits of the method, a comparative study of other related works was carried out. In another study, Nozari et al. [156] explored fault detection and isolation on the IPGT by using MLPs and a linear neuro-fuzzy method.

### 3.1.3   Black-box models of aero GTs

A nonlinear autoregressive moving average with exogenous inputs (NARMAX) model of an aircraft GT was estimated by Chiras et al. [157]. They employed nonparametric analysis in time and frequency domains to determine the order and nature of the nonlinearity of the system. The researchers combined time-domain NARMAX modeling, time and frequency domain analysis, identification techniques, and periodic test signals to improve GT nonlinear modeling. In another investigation, Chiras et al. [158] applied a forward-regression orthogonal estimation algorithm to make an NARMAX model for a twin-shaft Rolls-Royce Spey aircraft GT. A nonlinear relationship between dynamics of the shaft rotational speed and the fuel flow rate was also explored and discussed in the study. To validate the model performance, the researchers examined static and dynamic behavior of the engine for small and large signal test. The results were satisfactory and could be matched with the results from another previously estimated model. In another effort, Chiras et al. [159] used FFNN to model the relationship between fuel flow rate and shaft rotational speed dynamics for a Spey GT engine. They showed the necessity of using a nonlinear model for modeling high-amplitude dynamics of GT engines. Chiras et al. [160] also recommended a global nonlinear model of GT dynamics using NARMAX Structures. They investigated both linear and nonlinear models of a twin-shaft Rolls-Royce Spey GT. Their suggestion for a global nonlinear model was based on the fact that linear models vary with operational points. They discussed a simple method for identification of an NARX model. The performance of this model was satisfactory for both small and high amplitude tests. However, due to inherent problems with discrete-time estimation and great variability of the model parameters, the physical interpretability of the model was lost.

Ruano et al. [161] presented nonlinear identification of shaft-speed dynamics for a Rolls-Royce Spey aircraft GT under normal operation. They used two different approaches including NARX models and NN models. The researchers realized that among the three different structures of NN including radial basis function (RBF), MLP, and B-spline, the latter delivered the best results. They employed genetic programming tool for NARMAX and B-spline models to determine the model structure.

Two different configurations of backpropagation neural networks (BPNN) were developed by Torella et al. [162] to study and simulate the effects of GT air system on an aero engine performance. For the first configuration, to improve the accuracy of the model, different network structures in terms of training methods and number of hidden layers were investigated for on-design simulation of a large turbofan engine. For the second configuration, the researchers derived a computer code to set up BPNNs for simulation of the air system operation; working with or without faults. The applied methodology was very useful when diagnostics and troubleshooting of the air system were investigated. The researchers discussed the problems, the most suitable solutions, and the obtained results. They emphasized that the BPNN training did not cover multiple faults as well as the influence of sensor noise and fault on the air system fault identification.

Breikin et al. [163] employed a genetic algorithm (GA) approach for dynamic modeling of aero GT engines for condition monitoring purpose during the engine cruise operation. They applied real engine data to the algorithm to estimate parameters of the linear reduced-order model. They compared the results of the approach with traditional modeling techniques used in industry and realized that GA affords flexibility in the choice of performance metrics.

Badihi et al. [164] applied ANNs to estimate the fuel flow injection function to the combustor chamber of a jet engine. They used

Simulink®-MATLAB to make a mathematical model and to generate required data for training a feedforward MLP neural network. They showed that the resulting ANN model had the capability to predict performance parameters of the engine accurately. Mohammadi et al. [165] used MLP NNs with dynamic processing units for detection of faults in a twin-shaft aero GT engines. They verified the capability of the trained network by conducting varieties of simulations. NNs were employed by Loboda et al. [166] for fault identification of an aero GT. Both MLP and radial basis networks were used and compared in terms of accuracy and computation time. The results showed that the RBF was a little more accurate than MLP, but it needed much more computation time. Tayarani-Bathaie et al. [167] investigated using a set of single-input and single-output (SISO) dynamic neural network (DNN)-based models for fault detection of an aircraft engine. They carried out various simulations to demonstrate the performance of the proposed fault detection scheme. Kulyk et al. [168] proposed a methodology for obtaining test and training data sets and formulating input parameters of a static NN for diagnosing aero GT engines and recognizing individual and multiple defects in the air-gas path units. They considered the operation of the engine in a wide range of modes.

## 3.2   Black-box approach in control system design

The NN controllers typically suffer performance degradation when dealing with unstable inverse models. Besides, the stability and robustness of the NN approaches are difficult to be analyzed. However, the NN controllers are widely known for their excellent reference tracking capability and their flexibility for implementation on various systems. Although PID controllers are still being widely used in control loops in the majority of industrial plants, their algorithm might be difficult to deal with in highly nonlinear and time-varying processes [21]. For these reasons, the learning-based control methodology such as a NN has been widely used in various industrial applications. There is a strong motivation for the development of a large number of schemes for ANN-based controllers due to their successful industrial applications [169]. ANN-based models for control systems can be trained using the data generated from a previously simulated model of the plant, or can be obtained directly from special open loop experiments performed on the plant [170]. It has been demonstrated that the input–output data sets of the system parameters obtained from a plant, which is controlled by a linear controller, can be reliably used for ANN training process [171].

Remarkable efforts have been made during recent decades to use NNs based controllers (Neurocontrol) for industrial systems. A survey in Science Direct and IEEE databases shows that the number of papers in the field of neurocontrol has increased significantly from 1990 to

2008 [172]. Agarwal [169] presented a systematic classification of various neurocontrols and showed that the neurocontrol studies are essentially different despite all their similarities. Hunt et al. [173] and Balakrishnan and Weil [174] also carried out a survey in this area in 1992 and 1996, respectively. Rowen and Housen [175] investigated GT airflow control for optimum heat recovery and its advantages at GT part-load conditions. They discussed performance and control flexibility of both single-shaft and twin-shaft GTs in industrial heat recovery applications and demonstrated the adaptability of GTs in meeting unique industrial process requirements. Hagan et al. [176,177] presented an overview of NNs and their applications to control systems. Their research covered different issues such as MLP NN for function approximation, the backpropagation algorithm for training MLPs, several techniques for improving generalization, as well as three different control architectures including model reference adaptive control, model predictive control, and feedback linearization control.

Investigation for the practical use of ANN to control complex and nonlinear systems was carried out by Nabney and Cressy [178]. They utilized multiple ANN controllers to maintain the level of thrust for aero GTs and to control system variables for a twin-shaft aircraft GT engine in desirable and safe operational regions. The main idea behind the research was to minimize fuel consumption and to increase the engine life. They aimed to improve the performance of control system by using the capability of ANN in nonlinear mapping instead of using varieties of linear controllers. They used MLP architecture with a single hidden layer to train the networks. The researchers applied a reference model as an input to the ANN controller. The results showed that performance of the applied ANN controller was better than conventional ones. However, they could not track the reference models as closely as they had expected.

Another effort was carried out by Dodd and Martin [179], more or less with the same objectives. They proposed an ANN-based adaptive technique to model and control an aero GT engine and to maintain thrust at a desired level while minimizing fuel consumption in the engine. They suggested a technique, which consequently could lead to maximizing thrust for a specified fuel, lowering the critical temperature of the turbine blades, and increasing the engine life. In their research, an FFNN with sigmoid activation function was utilized to model the system. The simplicity and differentiability of the NN helped the researchers to calculate necessary changes to controllable parameters of the engine, and consequently, to maintain the level of the thrust in a targeted point. Figure 3.2 shows the block diagram of the ANN model. The inputs correspond to the fuel rate, final nozzle area, and inlet guide vane angle. The only output is thrust [179].

*Figure 3.2* **(See color insert.)** Block diagram of an ANN-based aero GT model for system optimization consists of minimizing fuel while maintaining thrust. (From N. Dodd and J. Martin, *Computing & Control Engineering Journal,* vol. 8, no. 3, pp. 129–135, 1997.)

Psaltis et al. [180] employed a multilayer NN processor and used different learning architectures to train the neural controller for a given plant. Lietzau and Kreiner [181,182] explored the principles and possible applications of model-based control concepts for jet engines. They investigated modeling methods for real-time simulation and online model adaptation. To improve the transient stability performance of a power system, Dash et al. [183] presented a radial basis function neural network (RBFNN) controller with both single and multineuron architecture for the unified power flow controller (UPFC). They observed that RBF controller with multineuron structure performed better and showed a superior damping performance compared with the existing PI controllers. They also demonstrated that RBF model is very useful for the purpose of real-time implementation.

Development of an intelligent optimal control system with learning generalization capabilities was explored by Becerikli et al. [184]. They used a DNN as a control trajectory priming system to overcome the nondynamic nature of popular ANN architectures. The trained DNN helped to generate the initial control policy close to the optimal result. Litt et al. [185] explored an adaptive, multivariable controller for deterioration compensation of the thrust due to aging in an aero engine GT. They used the relationship between the level of engine degradation and the overshoot in engine temperature ratio, which was the cause of the thrust response variation, to adapt the controller. A mathematical model of a combined cycle gas turbine (CCGT), as part of a large-scale national power generation network, was developed by Lalor and O'Malley [186]. The objective was to study the response of CCGT to the frequency disturbance and to investigate the effects of increasing proportions of CCGT generation on the entire network when the model was integrated into a larger model.

Junghui and Tien-Chih [21] presented a new control approach by employing a PID controller and a linearized neural network model. Their research objective was to make a balance between nonlinear and conventional linear control designs in order to improve the control performance for the nonlinear systems. Although the proposed method provided a useful physical interpretation of the system dynamics, and it was effective in reducing the variance of the system output caused by disturbances, there were several drawbacks such as convergence problems that could have a serious impact on the controller design.

Some of the researchers tried to develop ANN-based MPC for control of processes. Sahin et al. [187] proposed a neural network approach for a nonlinear model predictive control (NMPC). They showed that the MPC can be effectively employed to control nonlinear industrial processes without linearization requirement. Ławryńczuk [188,189] discussed details of NMPC algorithms for MIMO processes modeled by means of neural networks of a feedforward structure. Jadlovská et al. [190] presented classical, and NARX approaches to design generalized predictive control (GPC) algorithm for a nonlinear system. They concluded that the intelligent neural GPC controller performance which used linearization techniques showed significant advantages over the conventional nonlinear predictive controller. Suarez et al. [170] developed a new predictive control scheme based on neural networks to linearize nonlinear dynamical systems. Cipriano [191] discussed implementation of fuzzy predictive control for power plants using nonlinear models based on fuzzy expert systems, and using fuzzy logic to characterize the objective function and the constraints. An NMPC for frequency and temperature control of a heavy-duty IPGT was developed by Kim et al. [192]. They showed that the proposed control system has superior performance to PID control in terms of responses to disturbances in electrical loads.

Ghorbani et al. [193,194] and Mu and Rees [195] explored applications of ANN-based MPC to GTs. Mu and Rees [195] investigated nonlinear modeling and control of a Rolls-Royce Spey aircraft GT. They used NARMAX and neural networks to identify the engine dynamics under different operational conditions. The researchers applied an approximate model predictive control (AMPC) to control shaft rotational speed. The results proved that the performance of the AMPC as a global nonlinear controller was much better than gain-scheduling PID controllers. AMPC showed optimal performance for both small and large random step changes as well as against disturbances and model mismatch. In another effort, Mu et al. [196] examined two different approaches to design a global nonlinear controller for an aircraft GT. They compared and discussed the properties of AMPC and NMPC. The results showed that both controllers provided good performance for the complete operational range. However, AMPC showed better performance against disturbances and uncertainties. Besides, AMPC could be gained analytically, required less computational time, and avoided local minima.

A combination of RNN and reinforcement learning (RL) was employed by Schaefer et al. [197] to control a GT for stable operation at high load. High system identification quality of RNN could facilitate the network training by using limited available data sets. Sisworahardjo et al. [198] presented a neural network controller for power plant MGTs. They applied both PI and ANN controllers to control voltage, speed, temperature, and power. They concluded that ANN-based controller had a better performance in terms of error measures.

Yamagami et al. [199] developed an optimal control system for the GTs of CCPPs as a result of development of the control systems for the entire power plant including steam turbines, waste heat recovery boilers, and auxiliary machines. Implementation of an MPC on a heavy-duty power plant GT was investigated by Ghorbani et al. [193,194]. They built a model of the system based on a mathematical procedure, and autoregressive with exogenous input (ARX) identification method. The research objective was to design a controller that could adjust the rotational speed of the shaft and exhaust gas temperature by the fuel flow rate and the position of IGV. The MPC controller showed superior performance to both PID controller and SPEEDTRONIC control system.

Bazazzadeh et al. [200] developed a mathematical model of a controller for an aero GT by using fuzzy logic and MLP-based neural network methods in Simulink-MATLAB environment. The neural networks were employed as an effective method to define the optimum fuzzy fuel functions. The resulting controller could successfully achieve the desired performance and stability.

Using PID and ANN controllers for a heavy-duty GT plant was investigated by Balamurugan et al. [201]. Their work was based on the GT mathematical model previously developed by Rowen [36]. They applied Ziegler-Nichols method to tune PID controller parameters. Besides, they trained an ANN controller using backpropagation method to control the speed of the GT. The simulation results showed that the ANN controller performed better than the PID controller. Figure 3.3 shows a comparison of GT plant response with PID and ANN controllers [201].

## 3.3 Final statement

In the area of black-box models, there are many different types of ANN architectures in terms of network topology, data flow, input types, and activation functions such as recurrent, RBF, and Hopfield networks. The ANN models can also be trained with varieties of algorithms such as backpropagation Levenberg-Marquardt algorithm, nongradient-based training methods, and genetic algorithm.

As it can be seen, each of the research activities in the field of modeling of GTs investigated the issue from a specific perspective and had its own limitation. For instance, Chiras et al. [157–160], Ruano et al. [161], and Torella et al. [162] concentrated on ANN-based modeling of aero GTs. They employed a variety of ANN-based techniques and approaches such

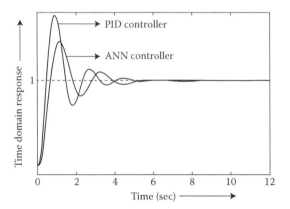

*Figure 3.3* A comparison of GT plant response with PID and ANN controllers. (From S. Balamurugan, R. J. Xavier, and A. Jeyakumar, *International Journal of Applied Engineering Research*, vol. 3, no. 12, pp. 1765–1771, 2008.)

as MLP, NARMAX, NARX, RBF, BPNN, and B-spline, to explore nonlinear dynamics of aero GTs.

While some researchers such as Jurado [136] and Bartolini et al. [137] investigated MGTs using ANN techniques, other researchers explored ANN-based IPGTs. The major contributions in this area include the studies carried out by Lazzaretto and Toffolo [139], Jurado [136], Bartolini et al. [137], Basso et al. [142], Bettocchi et al. [144,145,202,203], Yoru et al. [148], Simani and Patton [147], Palmé et al. [154], Fast et al. [149–151], Fast and Palmé [152], Fast [153], Spina and Venturini [146], Ogaji et al. [140], and Arriagada et al. [141]. The results of these studies have indicated that ANN can be very useful for the real-time simulation of GTs, specifically when there is not enough information about the system dynamics. It has also been shown that ANN could be used as a high-speed, powerful tool for real-time control problems [140]. ANN has the capability to identify system faults and to generate warnings at early stages with high reliability [141].

Ogaji et al. [140], Arriagada et al. [141], Fast et al. [149–151], Palmé et al. [154], Fast et al. [152], and Fast [153], explored applications of ANN for fault diagnosis, condition monitoring, and/or sensor validation purposes. Fast et al. [151] just considered a full-load situation for ANN-based system identification and modeling of single-shaft GTs. In some GT models, the nonlinear terms in the model were restricted to the second order [136]. Besides, most of the ANN-based models of GTs were built on the basis of a specific training function (*trainlm*) and transfer functions ("tansig" or "logsig" type in the hidden layer, and "purelin" type in the output layer). Besides, some research activities just concentrated on the dynamic behavior of individual main components of GTs such as compressors and combustors. For instance, one can refer to NN techniques employed by Ghorbani et al. [204,205], Palmé et al. [206], Mozafari et al. [207], and Sethi et al. [208].

As it can be seen, none of the past ANN research activities on GTs conducted an extensive performance comparative study using combinations of different network architectures, training algorithms, and different number of neurons. A comprehensive and comparative study in this field can be very useful in system identification and modeling of GT engines. Approximating an ANN model with high generalization, capabilities, and robustness for IPGTs can be extensively investigated using simulated data or operational data of real GTs and based on the flexibility that ANN provides for modeling of different types of systems. For this purpose, different ANN architectures can be explored for GTs in order to attain such a model that can predict dynamic behavior of the system as accurately as possible and can also be employed as a powerful tool in condition monitoring, troubleshooting, and maintenance of GTs.

## 3.4   Summary

This chapter presented a comprehensive overview of research activities in the field of black-box modeling, simulation, and control of GTs. It discussed most relevant scientific sources and significant research activities in this area for different kinds of GTs. Main black-box models and their applications to control systems were investigated for low-power aero, and power plant GTs. It was shown that further research needs to be carried out to resolve unpredictable challenges that arise in the manufacturing processes or in the operation of industrial plants.

# chapter four

# ANN-based system identification for industrial systems

> All things are artificial, for nature is the Art of God.
>
> **Sir Thomas Browne**
> *English Author, 1605–1682*

Ever since ANN was presented for the first time by Bernard Widrow from Stanford University in 1950s, it has been a constant challenge for researchers to find optimal ANN-based solutions to design, manufacture, develop, and operate new generations of industrial systems as efficiently, reliably, and durably as possible. Getting enough information about the system that is to be modeled is the first step in system identification and modeling process. Besides, a clear statement of the modeling objectives is necessary for making an efficient model. Industrial systems may be modeled for condition monitoring, fault detection and diagnosis, sensor validation, system identification or design, and optimization of control systems [209].

A variety of analytical and experimental methods has been suggested so far for the industrial system modeling. One of the novel approaches for system identification, and modeling of GTs is employing ANN-based techniques. ANN has the power to solve many complex problems; it can be used for function fitting, approximation, pattern recognition, clustering, image matching, classification, feature extraction, noise reduction, extrapolation (based on historical data), and dynamic modeling and prediction.

This chapter briefly presents ANNs and their main elements and structures. Next, ANN-based model building process including system analysis, data acquisition and preparation, network architecture, as well as network training and validation is explained. Different challenges in using ANN-based methodologies for industrial systems and their applications, advantages, and limitations are also discussed in this chapter.

## 4.1 Artificial neural network (ANN)

The main idea behind the creation of ANN was to resemble the human brain in order to solve complicated problems in a variety of scientific areas such as engineering, psychology, linguistics, philosophy, economics, neuroscience, and so on. ANN is defined as a computing system that is made up

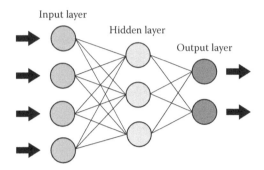

*Figure 4.1* **(See color insert.)** A simple structure of an ANN with input, hidden, and output layers.

of a group of simple highly interconnected processing elements (neurons) with linear or nonlinear transfer functions. These elements process information by their dynamic state response to external inputs [210]. Neurons are arranged in different layers including input layer, hidden layer(s), and output layer. The number of neurons and layers in an ANN model depend on the degree of complexity of the system dynamics. ANN learns the relation between inputs and outputs of the system through an iterative process called training. Each input into the neuron has its own associated weight. Weights are adjustable numbers, which are determined during the training process. Selecting the right parameters as inputs and outputs of ANN is very important for making an accurate and reliable model. The availability of data for the selected parameters, system knowledge for identification of interconnections between different parameters and the objectives for making a model are basic factors in choosing appropriate inputs and outputs. Accuracy of the selected output parameters can be examined by sensitivity analysis. Figure 4.1 shows a simple structure of a typical ANN with four inputs, two outputs, and three neurons in one hidden layer.

## 4.2   The model of an artificial neuron

Artificial neuron is the basic and fundamental element of all ANN structures. Figure 4.2 shows a simple single-input neuron with its input, output, and components including the sum and function blocks [211].

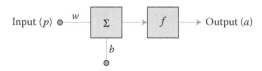

*Figure 4.2* Single-input neuron structure.

$p$, $w$, $b$, $f$, and $a$, are scalar input, scalar weight, bias, transfer (activation) function, and scalar output, respectively. The neuron output is calculated by Equation 4.1.

$$a = f(w^*p + b) \tag{4.1}$$

The parameters $w$ and $b$ can be adjusted by learning rules so that the relationship between the input and output meets the expected goal [176]. Bias is a weight that is not connected to other nodes, and its input is always set to one. The purpose of bias is to offset the origin of the transfer function for more rapid convergence. Thus, bias allows a node to have an output even if the input is zero.

A neuron usually has more than one input. Figures 4.3 and 4.4 show multiple-input neuron structures with one and multiple neurons in the hidden layer, respectively [176]. $R$ and $S$ indicate the number of elements in input vector, and the number of neurons in the layer. In this case, the input $P$, the weight $w$, and the output $a$, would be vectors, and Equation 4.1 would have a matrix nature as it is shown in Figures 4.3 and 4.4. An NN may have several layers operating in parallel. Each layer has its own inputs, outputs, and components.

## 4.3  ANN-based model building procedure

ANN, as a data-driven model, has been considered as a suitable alternative to white-box models during the last few decades. ANN models for GTs can be created using different approaches due to the varieties of network structures, training algorithms, type of the activation functions, number of neurons, number of hidden layers, values of weights and biases, as well as data structures. However, the best structure for ANN is the one that can predict the dynamic behavior of the system as accurately

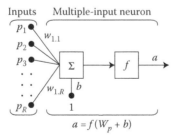

*Figure 4.3* Multiple-input neuron structure. (From M. T. Hagan, H. B. Demuth, and O. D. JESÚS, *International Journal of Robust and Nonlinear Control*, vol. 12, no. 11, pp. 959–985, 2002.)

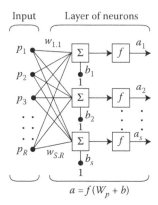

**Figure 4.4** Matrix form of multiple-input neuron structure. (From M. T. Hagan, H. B. Demuth, and O. D. JESÚS, *International Journal of Robust and Nonlinear Control*, vol. 12, no. 11, pp. 959–985, 2002.)

as possible. The following presents the main steps for setting up an ANN-based model.

### 4.3.1   System analysis

Before training of any ANN model, it is normally needed to do an extensive system study including system configuration and history record, technical characteristics, operational conditions, monitoring system, available parameters, sensors situation and reliability, accessibility of the system data, availability of performance curves, and so on. This step is necessary to establish a suitable input and output structure for the ANN model. It is also necessary to find out the method that is more compatible with the research expectations.

### 4.3.2   Data acquisition and preparation

Data acquisition is the first step and a vital part of ANN-based modeling and control of an industrial system. ANN-based models can be created directly using the operational data from an actual GT available in a variety of industrial power plants. The data can be obtained offline, if the system is run in idle mode. However, in this case, the effect of load changes may not be investigated.

When operational data are not available, simulated data from original equipment manufacturers (OEMs) performance or generated by engineering and/or commercial software such as Simulink®-MATLAB® may be used. In the latter case, system information is fed to the software to make a preliminary model for data generation and to set up the black-box model.

The obtained data should cover the complete operational range of the system. All transient data during start or stop processes should be removed from the collected data before the modeling process. The required data for modeling of IPGTs can be collected from the GTs available in a variety of industrial power plants all over the world.

Format of the data structures affects network training. Input vectors can occur concurrently or sequentially in time. For current vectors, inputs occur at the same time or in no particular time sequence. In this case, order is not important and a number of networks can run in parallel. One input vector can be presented to each of the networks. For concurrent vectors, inputs occur sequentially in time, and the order in which the vectors appear is important. ANN Toolbox in MATLAB can be employed to model the system and to design the appropriate control system after the stage of data acquisition is completed.

### 4.3.3   Network architecture

ANN can be classified into two different categories, static (feedforward) and dynamic (feedback) networks. In static networks, there is no feedback element or delay, and output can be calculated directly from the input through feedforward connections. In dynamic networks, the output depends both on the current input to the network and on the current or previous inputs, outputs, or states of the network. Figure 4.5 shows an FFNN with three layers [212]. Figure 4.6 shows an NARX network with two layers [212].

#### 4.3.3.1   Feedforward neural network

As Figure 4.5 shows, neurons in an FFNN model are grouped into layers, which are connected to the direction of the passing signal (from left to right, in this case). There are no lateral connections within each layer and no feedback connections within the network. The best-known ANN of this type is MLP [212]. There is at least one hidden layer in an FFNN. MLP is one of the most commonly used ANN in scientific applications. It can be used for function fitting, pattern recognition, and nonlinear classification. Among the different ANN structures, MLP is the first choice for modeling and simulation of nonlinear behavior of industrial systems such as GTs [213].

#### 4.3.3.2   Feedback neural network

Feedback NN, also called dynamic or recurrent NN, is a type of ANN structure that allows modeling of time-domain behaviors of a dynamic system. The outputs of a dynamic system depend not only on present inputs, but also on the history of the system states and inputs. A recurrent NN structure is needed to model such behaviors.

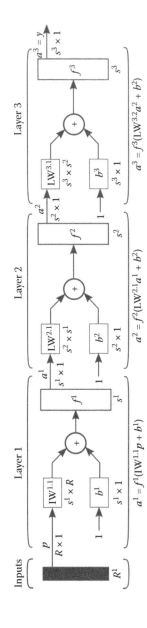

***Figure 4.5*** Three-layer FFNN. (From M. H. Beale, M. T. Hagan, and H. B. Demuth, Neural Network Toolbox™ User's Guide, R2011b ed., Natick, MA: MathWorks, 2011, 404 pp.)

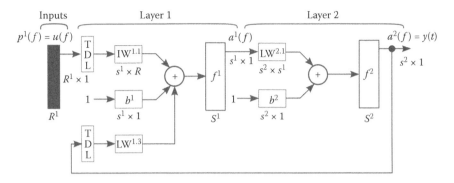

*Figure 4.6* NARX network with a two-layer feedforward network. (From M. H. Beale, M. T. Hagan, and H. B. Demuth, Neural Network Toolbox™ User's Guide, R2011b ed., Natick, MA: MathWorks, 2011, 404 pp.)

One of the most commonly used feedback NNs is NARX. It is a recurrent network with feedback connections enclosing several layers of the network. NARX network has many applications; it can be used for modeling of nonlinear dynamic systems such as IPGTs, and can also be employed for nonlinear filtering purposes to make the target output as a noise-free version of the input signal. As Figure 4.6 shows, NARX model can be implemented by using an FFNN to approximate the function $f$ [212]. In this figure, a two-layer feedforward network is used for the approximation. The dependent output signal $y(t)$ is regressed on previous values of the output signal and previous values of an independent (exogenous) input signal. *TDL* indicates the time delay [212].

## 4.3.4 *Network training and validation*

Training or learning paradigms for an ANN can be mainly classified as supervised and unsupervised. In supervised learning, inputs and targets (desired outputs) are known, and the ANN model is trained in a way that maps inputs to the outputs. Supervised learning is employed for regression and classification purposes. However, in an unsupervised learning, targets are unknown, and the underlying relationship within the data sets has to be disclosed by the ANN using the data clustering method. Unsupervised learning is used for filtering and clustering of data.

There are two different styles of training in ANNs; incremental training and batch training. In incremental training, weights and biases of the network are updated each time an input is presented to the network. In batch training, the weights and biases are only updated after all inputs are presented. Batch training methods are generally more efficient in MATLAB environment, and they are emphasized in the NN

Toolbox software. However, there are some applications where incremental training can be useful so that the paradigm is implemented as well [212].

The training process of ANN involves the variation of one or more parameters. For example, it is needed to change the number of neurons in the hidden layer in order to attain the best converging network. The number of neurons indicates the complexity that can be approximated by the NN. It is desirable to use the simplest possible network structure with the least number of input parameters. The developed model can be utilized to validate new process measurements. A true NN training procedure is usually based on an iterative approximation in which the parameters are successively updated in numerous steps. Such a step can be based on a single data item, on a set of them, or on all available data points. In each step, the desired outcome is compared with the actual one, and using the knowledge of the architecture, all parameters are changed slightly such that the error for the presented data points decreases [212].

Before training is started, the collected data are divided into three subsets including training, validation, and test data sets. The first subset is the training set, which is used for computing the gradient and updating the network weights and biases. The second subset is the validation set which is used to verify the model that has been created. The error on the validation set is monitored during the training process. The validation error decreases during the initial phase of training, as does the training set error. The network weights and biases are saved at the minimum of the validation set error. And finally, the test set is used after training and validation for a final test. Testing the NN with similar data as that used in the training set is one of the few methods used to verify that the network has adequately learned the input domain. In most instances, such testing techniques prove adequate for the acceptance of an NN system.

The validation data set is used to stop training early, if further training on the primary data will hurt generalization of the validation data. Test vector performance can be used to measure how well the network generalizes beyond primary and validation data. When the training is complete, the network performance can be checked to see if any changes need to be made to the training process, the network architecture, or the data sets. The first thing to do is to check the training *record;tr*. This structure contains all the information concerning the training of the network. For example, *tr.trainInd*, *tr.valInd*, and *tr.testInd* contain the indices of the data points that were used in the training, validation, and test sets, respectively. The *tr* structure also keeps track of several variables during the course of training such as the value of the performance function, the magnitude of the gradient, and so on.

### 4.3.4.1 Number of hidden layers and neurons

Choosing the right number of hidden layers and available neurons in each layer is very vital in training an NN. It has already been shown that any multidimensional nonlinear mapping of any continuous function can be carried out by a two-layer MLP with suitably chosen number of neurons in its hidden layer [214]. Therefore, the main task in modeling industrial systems using MLP is to determine the right number of neurons in the hidden layer for approaching an optimal ANN. Although increasing the number of neurons sometimes is necessary to catch nonlinear dynamics of the system, it does not mean that it can always and necessarily improve the model accuracy and generalizability.

### 4.3.4.2 Training algorithms

Different training algorithms can be used for training ANNs. The available training algorithms in the NNT software, which use gradient or Jacobian-based methods are as follows [212]:

- *Trainlm:* Levenberg-Marquardt
- *Trainbr:* Bayesian Regularization
- *Trainbfg:* BFGS Quasi-Newton
- *Trainrp:* Resilient Backpropagation
- *Trainscg:* Scaled Conjugate Gradient
- *Traincgb:* Conjugate Gradient with Powell/Beale Restarts
- *Traincgf:* Fletcher-Powell Conjugate Gradient
- *Traincgp:* Polak-Ribiére Conjugate Gradient
- *Trainoss:* One Step Secant
- *Traingdx:* Variable Learning Rate Gradient Descent

### 4.3.4.3 Transfer functions

Transfer (activation) functions transform activation level of a unit (neuron) into an output signal [215]. There are various transfer functions included in the NN Toolbox software. Karlik et al. [215] and Debes et al. [216] discussed various transfer functions and their applications to NNs. The two common transfer functions that are employed for MLP are *Log-Sigmoid* and *Tan-Sigmoid*. These functions are differentiable and can cope with nonlinearity of the industrial systems. Figure 4.7 shows different transfer functions that can be used for training NNs.

### 4.3.4.4 Weight values

Before training an ANN, the initial values of weights and biases have to be determined. Initialization of the weights and biases can be done automatically by the ANN Toolbox software, or it can be adjusted manually through writing and running codes in MATLAB.

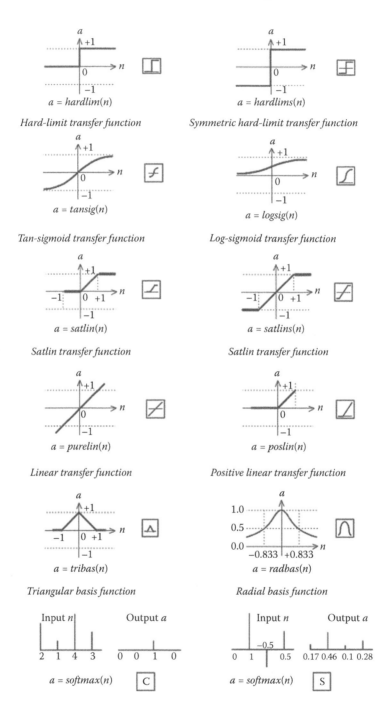

*Figure 4.7* NN transfer functions.

### 4.3.4.5   Error criteria

The objective of training an NN is to minimize the error as much as possible. Minimization of error simply means improving performance of the training and getting a more accurate model. Different definitions and types of error may be considered during training an NN. For instance, absolute error is defined as the difference between the measured (actual) output and the desired output (target). However, it is common to use mean square error (*MSE*) or root mean square error (*RMSE*) when training MLPs. *MSE* and *RMSE* are defined according to Equations 4.2 and 4.3 respectively, where $y_m$ is the measured data, $y$ is the prediction of the model, and $n_d$ is the number of data sets. Minimization of the error can be achieved by changing weights and/or training algorithms.

$$MSE = \frac{1}{n_d} \sum_{i=1}^{n_d} \left( \frac{y_{mi} - y_i}{y_{mi}} \right)^2 \tag{4.2}$$

$$RMSE = \sqrt{\frac{1}{n_d} \sum_{i=1}^{n_d} \left( \frac{y_{mi} - y_i}{y_{mi}} \right)^2} \tag{4.3}$$

### 4.3.4.6   Training stop criteria and overfitting

When training an NN, a stop criterion is determined to avoid what is called overtraining or overfitting. ANN has the potential tendency to overfit during the training process. Overfitting can occur during the training process when the ANN gets too specialized to fit the training data extremely well but at the expense of reasonably fitting the validation data. Overfitting is reflected by the steady increase in the validation error accompanied by a concomitant steady decrease in the training error. Poor performance due to overfitting is one of the most common problems in training ANNs. It can be overcome by using the cross-validation method, decreasing the number of neurons in a hidden layer(s), or adding a penalty term to the objective function for large weights. By using the cross-validation method, the network performance is measured during the training process, and if any incentive is given, the training is stopped before maximum number of epochs is reached. Epoch is an NN term for iteration in a training process. The number of epochs shows the number of times that all patterns are presented to the NN. More epochs mean more training time. In each epoch of an ANN, all the weight values of the neurons are updated.

## 4.4    *ANN applications to industrial systems*

ANN as an emerging research area has a wide range of potential applications that spans over science, art, engineering, and so on [217]. It has many advantages over conventional modeling approaches [218,219]. These advantages are due to the special structure and algorithm of the network. ANN methodology can be a suitable alternative to classical statistical modeling techniques when obtained data sets indicate nonlinearities in the system [220,221]. It has a demonstrated capability to solve combinatorial optimization problems in industrial plants [222].

ANN is a powerful tool for system identification and modeling due to its excellent ability to approximate uncertain nonlinearity to a high degree of accuracy. It can perform implicit nonlinear modeling and filtering of the system data [222] and detect coupled nonlinear relations between independent and dependent variables without any need for dynamic equations [223,224]. ANN offers a cost-effective and reliable approach to condition monitoring. The collected data related to the condition of the system can be classified and trained by using ANNs in order to generalize a methodology for data analysis at any time of the measurement. ANN can be applied to examine condition-based maintenance [149] to detect anomalies [151] and to isolate faults [147] in the performance of industrial systems. Using ANN for sensor validation leads to more cost-effective maintenance. ANN-based methodology can be developed to minimize the need for calibration of sensors and to decrease the percentage of shutdowns due to sensor failure [154].

ANN has been considered as an acceptable solution to many outstanding problems in modeling and control of nonlinear systems. Real data obtained from an industrial system can be used to develop a simple ANN model of the system with very high prediction accuracy [150]. In control design process, an NN may directly implement the controller (direct design). In this case, an NN will be trained as a controller based on some specified criteria. It is also possible to design a conventional controller for an available ANN model (indirect design). In this case, the controller is not itself an NN. In many cases, the obtained data from the systems located in industrial factories and plants may include noisy data. Besides, some sorts of data may be inaccurate or incomplete due to faulty sensors. These tend to happen when the system is old, and/or maintenance is poor. ANN has the capability to work considerably well even when the data sets are noisy or incomplete. It can learn from incomplete and noisy data [225].

The development of ANN requires less formal statistical training [226]. Training ANN is simple and does not need professional statistical knowledge. If the data sets and appropriate software are available, then even newcomers to the field can handle the training process. However, experience and statistical background can still be very useful and effective during the whole process. ANN can be developed using different training algorithms [226]. It also has the capability of dealing with stochastic variations of the scheduled operating point with increasing data and can be used for online processing and classification [222].

In addition to applications of ANN to industrial systems, it has many general advantages such as simple processing elements, fast processing time, easy training process, and high computational speed. Capturing any kind of relation and association, exploring regularities within a set of patterns, and having the capability to be used for very large number and diversity of data and variables, are other characteristics of ANN. It provides a high degree of adaptive interconnections between elements and can be used where the relations between different parameters of the system are difficult to uncover with conventional approaches. ANN is not restricted by variety of assumptions such as linearity, normality, and variable independence, as many conventional techniques are. It can even generalize the situations for which it has not been previously trained. Generally, it is believed that the ability of ANN to model different kinds of industrial systems in a variety of applications can decrease the required time on model development, and thus, leads to a better performance compared with conventional techniques [18].

## 4.5 ANN limitations

Like any modeling technique, ANN has its own limitations based on the particular application and methodology under consideration [218,219]. The basic challenges to be resolved include training time, upgrading of trained neural nets, selection of the training vector, and integration of technologies in the problem domain [222]. Despite all investigations carried out so far, ANN as a black-box technique is still restricted to clearly identify the importance of every single input parameter during the training process [226]. There is little intuitive information about what actually happens inside the network during the learning process, and one can hardly intuitively interpret the internal workings of an ANN. There are many issues in terms of methodology, which need to be resolved [226].

There are remarkable difficulties in using ANN models on industrial sites. Compared with other conventional models, ANN models may be more difficult to use on operational fields. Special software and hardware are required to implement the model. The correct interpretation of the output is not easy either. To implement an ANN model, sometimes many computational resources such as mainframes, minicomputers, and processors are needed. For more complicated systems, more resources are required [226].

To train, validate, and test an ANN, usually, a large number of data sets are required. There is no fixed number of data sets for an optimal training process. It may differ case by case for different industrial systems. However, the amount of the obtained data should be large enough to disclose underlying structure of the system as accurately as possible and to provide sufficient understanding of the system dynamics. The data sets may be on-site operational data or simulated data by a previously confirmed model. Data acquisition especially on operational sites may be a difficult and time-consuming process. New data sets cannot be fed directly to the trained ANN to improve its performance, and it is needed to be trained again against all the available data sets. Manipulating time-series data in ANN is also a complicated issue. A unique ANN model is trained to solve just a specific problem. It means that getting good results from an ANN model for a specific problem does not guarantee solutions to other problems. ANN relies on empirical development. The relatively new technique still needs to be developed based on the practical implementations and experiments gained by researchers.

## 4.6   Summary

There are different approaches and methodologies in system identification and modeling of industrial systems. ANN is increasingly considered as a suitable alternative to white-box models over the last few decades. The nature and strength of the interrelations of system variables, as well as the nature of applications, are vital criteria for training an NN with sufficiently rich empirical data.

This chapter briefly introduced ANNs and their types and structures. It also provided details of an ANN-based model building procedure including system analysis, data acquisition and preparation, network architecture, as well as network training and validation. Applications and limitations of ANN approach for system identification and modeling were also discussed in this chapter. It is important to notice that approximation and error are inseparable parts of any system identification method, and ANN is not an exception. Many issues should be considered when a comparison is made between ANN and any of the conventional modeling techniques.

*chapter five*

# Modeling and simulation of a single-shaft GT

Imagination is more important than knowledge.

**Albert Einstein**
*German-American Physicist, 1879–1955*

GTs have been used widely in industrial plants all over the world. They are the main source of power generation in places such as offshore plants and oil fields, which are far away from urban areas. The key role of GTs in the developing industry has motivated researchers to explore new methodologies in order to be able to predict the dynamic behavior of these complex systems as accurately as possible. A variety of analytical and experimental techniques has been developed so far to approach an optimal model of GTs. Fortunately, black-box system identification techniques and specifically ANN-based approaches can effectively assist researchers who work in this field. The study in this area can be categorized into IPGT, aero, and low-power GT models [227,228]. ANN is one of the techniques that has played a significant role in system identification and modeling of industrial systems. This is due to its capability to capture dynamics of the systems without any prior knowledge about their complicated dynamical equations. Because of sophisticated and nonlinear dynamic behavior of GTs, significant attention still needs to be paid to the dynamics of these systems to unfold unknowns behind undesirable events during GT operation. As it can be seen, each research activity in the field of modeling of GTs investigated the issue from a specific perspective and had its own limitation. According to the methodology used in this chapter, various backpropagation training functions, different number of neurons, and a variety of transfer functions were employed to train the network in order to explore an accurate ANN model using an MLP structure. To increase the level of generalization for the model, the data sets were partitioned randomly for training, validation, and test purposes.

In this chapter, first a Simulink® model of a low-power GT based on the research [27] is presented. Then, an ANN-based system identification process is developed. The process includes generating the required data sets from the Simulink model, writing the computer program code, and training the network. Finally, the results are presented, and concluding remarks are discussed [229].

## 5.1 GT Simulink® model

The data used in this chapter was generated using a resimulated nonlinear dynamic model of a low-power single-shaft GT. The model has been already developed and verified for loop-shaping control purposes by Ailer et al. [27]. The main idea of their research was to improve dynamic response of the engine by implementation of a developed nonlinear controller. The model was developed and simulated in Simulink-MATLAB®, based on engineering principles, GT dynamics, constitutive algebraic equations, and by using operational data. Model verification was performed by open-loop simulations against qualitative operation experience and engineering intuition [27]. Figure 5.1 shows a schematic of the main components of a single-shaft GT engine including compressor, combustion chamber (combustor), and turbine.

The Simulink model was built using the same principles and equations. Equations 5.1 through 5.3 are the main equations of the GT employed in the Simulink model [27]. Definition of each of the parameters in these equations is provided in Table 5.1. A simplified feature of the Simulink model is shown in Figure 5.2. In this figure, $N$ and $T_{04}$ are shown as the outputs of the system, $\dot{m}_f$, $M_{load}$, $T_{01}$, and $P_{01}$ are considered as inputs of the model. The other GT parameters can also be considered as outputs of the system. Figures 5.3 through 5.8 show subsystems of the Simulink model in a MATLAB environment.

$$\frac{dm_{cc}}{dt} = \dot{m}_c + \dot{m}_f - \dot{m}_t \tag{5.1}$$

$$\frac{dP_{03}}{dt} = \frac{P_{03}}{m_{cc}}(\dot{m}_c + \dot{m}_f - \dot{m}_t) + \frac{P_{03}}{T_{03} \, C_{v_{med}} m_{cc}} \tag{5.2}$$
$$(\dot{m}_c C_{p_{air}} T_{02} - \dot{m}_t C_{p_{gas}} T_{03} + q_f \eta_{cc} \dot{m}_f - C_{v_{med}} T_{03}(\dot{m}_c + \dot{m}_f - \dot{m}_t))$$

$$\frac{dN}{dt} = \frac{1}{4\pi^2 In}\left( \dot{m}_t C_{p_{gas}}(T_{03} - T_{04})\eta_{mech} - \dot{m}_c C_{p_{air}}(T_{02} - T_{01}) - 2\pi \frac{3}{50} N M_{load} \right) \tag{5.3}$$

## 5.2 ANN-based system identification

During recent decades, ANN-based models have been considered as suitable alternatives to white-box models. In this section, ANN-based system identification for the GT is carried out through data generation and training processes.

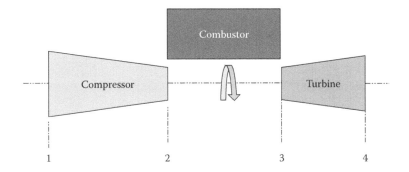

*Figure 5.1* A schematic of a typical single-shaft GT.

*Table 5.1* Definition of parameters in Equations 5.1 through 5.3

| Parameter | Symbol | Unit |
|---|---|---|
| Rotational speed (number of revolutions) | $N$ | 1/s |
| Temperature at section 1 | $T_{01}$ | K |
| Temperature at section 2 | $T_{02}$ | K |
| Temperature at section 3 | $T_{03}$ | K |
| Temperature at section 4 | $T_{04}$ | K |
| Pressure at section 1 | $P_{01}$ | Pa |
| Pressure at section 2 | $P_{02}$ | Pa |
| Pressure at section 3 | $P_{03}$ | Pa |
| Pressure at section 4 | $P_{04}$ | Pa |
| Air mass flow rate in compressor | $\dot{m}_c$ | kg/s |
| Gas mass flow rate in turbine | $\dot{m}_t$ | kg/s |
| Fuel mass flow rate | $\dot{m}_f$ | kg/s |
| Gas mass in the combustion chamber | $\dot{m}_{cc}$ | kg |
| Time | $T$ | s |
| Specific heat of air in constant pressure | $C_{pair}$ | J/kg K |
| Specific heat of gas in constant pressure | $C_{pgas}$ | J/kg K |
| Medium specific heat in constant volume | $C_{vmed}$ | J/kg K |
| Lower thermal value of fuel | $q_f$ | J/kg |
| Combustion chamber efficiency | $\eta_{comb}$ | — |
| Mechanical efficiency | $\eta_{mech}$ | — |
| Moment of inertia | $I$ | kg m² |
| Moment of load | $M_{load}$ | N m |

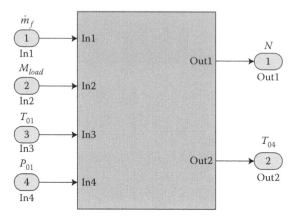

**Figure 5.2** Simplified Simulink® model of the GT.

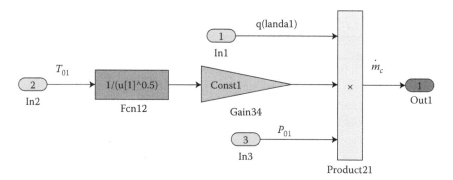

**Figure 5.3** Subsystem number 1 of the GT Simulink® model for creating mass flow rate in the compressor.

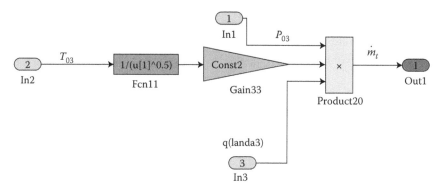

**Figure 5.4** Subsystem number 2 of the GT Simulink® model for creating mass flow rate in the turbine.

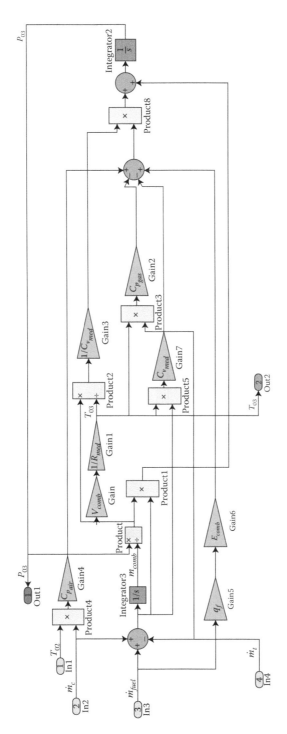

*Figure 5.5* Subsystem number 3 of the GT Simulink® model for creating $T_{03}$ and $P_{03}$.

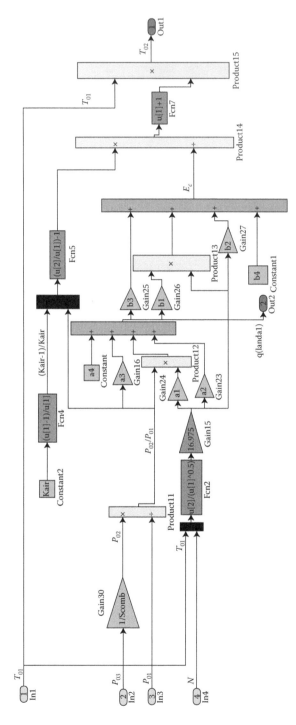

***Figure 5.6*** Subsystem number 4 of the GT Simulink® model for creating $T_{02}$.

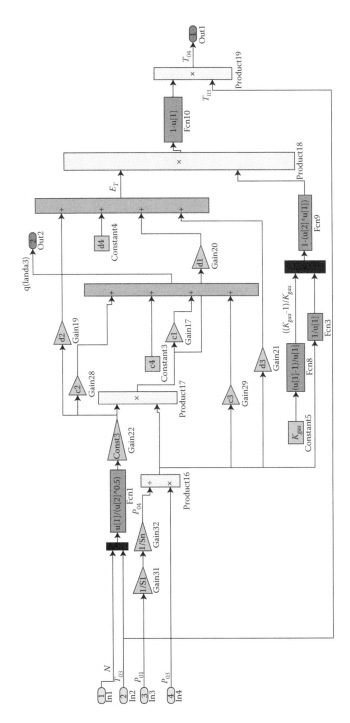

*Figure 5.7* Subsystem number 5 of the GT Simulink® model for creating $T_{04}$.

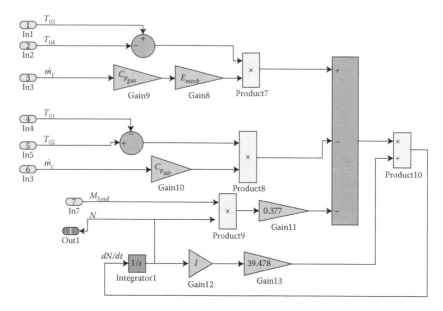

**Figure 5.8** Subsystem number 6 of the GT Simulink® model for creating rotational speed equation.

## 5.2.1　Data generation

The required data for the ANN-based modeling were generated for the entire operational range of the engine using the Simulink model already explained in this chapter. About 3000 such data sets were employed for training an accurate ANN-based model for the purpose of system identification. In this work, four variables including fuel rate, compressor inlet temperature and pressure, and the moment of load were considered as inputs. The outputs of the model consist of 17 different GT parameters. Tables 5.2 and 5.3 show the inputs and outputs (targets) of the model. Indices 1, 2, 3, and 4 refer to the corresponding sections of the single-shaft GT shown in Figure 5.1. It is necessary to say that the generated data sets are not of time-series type and do not show a continuous dynamics of the

**Table 5.2** GT input parameters for the ANN-based model

| Parameter | Symbol | Unit | Operational range |
|---|---|---|---|
| Fuel mass flow rate | $\dot{m}_f$ | kg/s | [0.00367; 0.027] |
| Compressor inlet temperature | $T_{01}$ | K | [243.15; 308.15] |
| Compressor inlet pressure | $P_{01}$ | kPa | [60; 110] |
| Moment of load | $M_{load}$ | N m | [0; 363] |

*Table 5.3* GT output parameters for the ANN-based model

| Parameter | Symbol | Unit |
|---|---|---|
| Rotational speed (number of revolutions) | $N$ | 1/s |
| Temperature at section 1 | $T_{01}$ | K |
| Temperature at section 2 | $T_{02}$ | K |
| Temperature at section 3 | $T_{03}$ | K |
| Temperature at section 4 | $T_{04}$ | K |
| Pressure at section 2 | $P_{02}$ | kPa |
| Pressure at section 3 | $P_{03}$ | kPa |
| Pressure at section 4 | $P_{04}$ | kPa |
| Air mass flow rate | $\dot{m}_{air}$ | kg/s |
| Compressor efficiency | $\eta_c$ | — |
| Turbine efficiency | $\eta_t$ | — |
| Compressor power | $\dot{W}_c$ | kW |
| Turbine power | $\dot{W}_t$ | kW |
| Net GT power | $\dot{W}_{net}$ | kW |
| GT efficiency | $\eta_{gt}$ | — |
| Specific fuel consumption | $SFC$ | kg/kWh |
| Mass ratio (flow rate) of fuel to air | $F$ | — |
| Pressure ratio in compressor | $PR_c$ | — |

system. They are used to train an MLP NN to predict the output parameters of the system based on the values of input parameters.

## 5.2.2 Training process

It has been already shown that any multidimensional nonlinear mapping of any continuous function can be carried out by a two-layer MLP with suitably chosen number of neurons in its hidden layer [214]. Therefore, an MLP model with two layers was employed for system identification of the GT. Figure 5.9 shows a schematic of the ANN structure of the GT. As it can be seen from this figure, the inputs and desired outputs correspond to the 4 and 17 GT parameters respectively. The ANN can be named 4-H-17 according to its structure with one hidden layer.

## 5.2.3 Code generation

To obtain an accurate network structure and to assure good generalization characteristic of the GT model, a comprehensive computer code was generated and run in MATLAB for a two-layer MLP network consisting of various backpropagation training functions, transfer functions, and different number of neurons. The 13 different training functions, applied

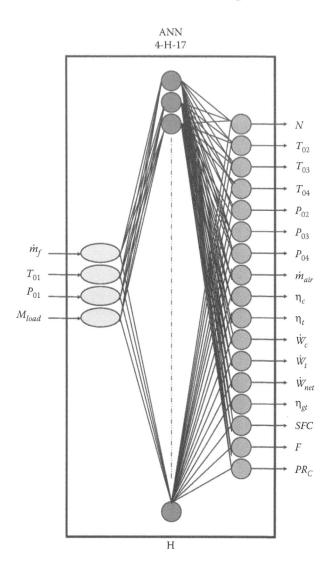

*Figure 5.9* **(See color insert.)** A schematic of the ANN structure for the GT engine.

in the code, included *trainbfg, trainb, traincgb, traincgf, traincgp, traingd,*
*traingda, traingdm, traingdx, trainlm, trainoss, trainrp,* and *trainscg.* The six
transfer functions employed in the code consisted of *tansig, logsig, purelin,*
*hardlim, satlin,* and *poslin.* The number of neurons tested in the program
varied from 1 to 40.

Figure 5.10 shows the flow diagram of the computer code for ANN-
based system identification of the GT. As it can be seen from Figure 5.10,

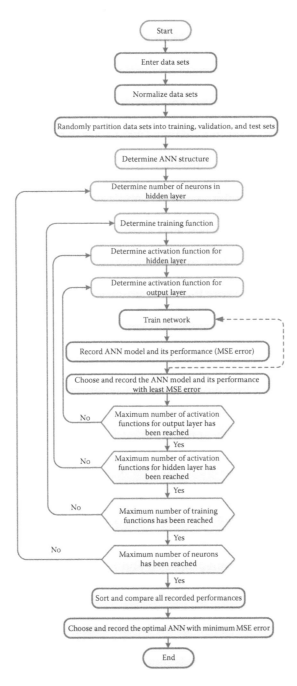

*Figure 5.10* **(See color insert.)** Flow diagram of the generated computer code for ANN-based system identification of the GT.

after feeding and normalizing the data sets, they are randomly partitioned into training (70%), validation (15%), and test (15%) categories. At
the next step, the structure of the network (MLP) is specified. After determination of number of neurons in the hidden layer, training function as
well as transfer functions for the hidden and output layers, training process of the network is started and repeated two more times for the same
adjusted factors so that the best performance among the three trials is
specified and recorded in a matrix. The process is repeated in four main
loops of the code for different number of neurons (1–40), different types of
backpropagation training functions, as well as combinations of different
transfer functions for the hidden and output layers. The results of all performances are recorded into a matrix and are sorted on the basis of their
performance errors *MSE*. According to the code, for each training process,
the best performance is calculated as an average of the best test performance and one-epoch-ahead of the best validation performance, which is
called average performance in this chapter.

One thousand epochs was considered for the entire training process of the ANN to be sure that the training would not be stopped before
reaching a dominating local minimum. Finally, the accurate ANN model
is recognized from the sorted results and tested again for verification.

## 5.3   Model selection process

In order to find the best model for the GT engine, the generated code was
run in MATLAB and 18720 (40*13*6*6) different ANN structures were
trained using randomly partitioned data sets for training, validation, and
test purposes. The results of the trainings were recorded, and the performances were evaluated and compared in terms of their *MSE*. Finally, the
most accurate MLP with minimum *MSE* was selected and tested again to
assure good generalization characteristics of the model. The results from
the model for different parameters of the GT (predicted values) were compared with the values of the generated data from the Simulink model. Table
5.4 indicates the best performances in terms of different training functions.
As it can be seen, a two-layer MLP with 20 neurons in the hidden layer,
using *trainlm* as its training function and *tansig* and *logsig* as its transfer
functions for the hidden and output layers showed the best performance. It
can also be seen that *trainlm* has a superior performance in terms of minimum *MSE*, compared with each of the other training functions.

Figure 5.11 shows details of the best resulting network based on the
average performance of all the trained structures. Performance of the ANN
for training, validation, and test is shown in this figure. As it can be seen, the
iteration in which the validation performance error reached the minimum
is 24. The MSE of the performance at this point is quite low. The training had
continued for 10 more iterations before the training stopped.

*Table 5.4* Best performance for different training functions

| Training function | Number of neurons | Transfer function in hidden layer | Transfer function in output layer | *MSE* for best average performance |
|---|---|---|---|---|
| *trainlm* | 20 | *Tansig* | *logsid* | 2.49E-06 |
| *traincpg* | 20 | *Purelin* | *hardlim* | 1.11E-05 |
| *traincgf* | 20 | *Logsig* | *logsig* | 1.15E-05 |
| *traincgb* | 18 | *Logsig* | *satlin* | 1.16E-05 |
| *trainscg* | 20 | *Tansig* | *satlin* | 1.36E-05 |
| *trainbfg* | 19 | *Purelin* | *satlin* | 1.50E-05 |
| *trainoss* | 20 | *Tansig* | *logsig* | 6.68E-05 |
| *trainrp* | 16 | *Tansig* | *satlin* | 0.000189 |
| *traingdx* | 15 | *tansig* | *satlin* | 0.000251 |
| *traingda* | 10 | *satlin* | *satlin* | 0.000688 |
| *trainb* | 4 | *hardlim* | *hardlim* | 0.063675 |
| *traingdm* | 2 | *tansig* | *logsig* | 0.087112 |
| *traingd* | 2 | *logsig* | *hardlim* | 0.090159 |

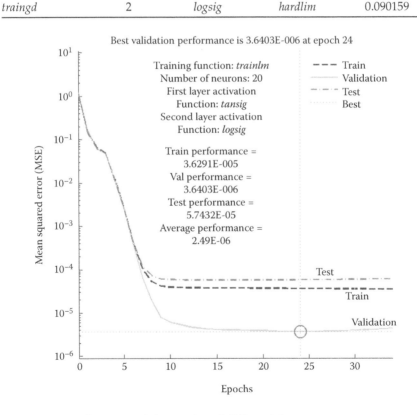

*Figure 5.11* Performance of the resulting MLP model.

Figure 5.12 shows the regression plot that indicates the relationship between outputs of the network and outputs of the system (targets). The R-value is an indication of the relationship between the outputs and the targets. As Figure 5.12 shows, R-values for all the graphs are very close to one. Therefore, the results of each of training, validation, and test data sets indicate a very good fit. Figures 5.13 through 5.28 compare output GT parameters of the Simulink and ANN-based models. For clarity of the figures, only outputs of 200 data sets out of 3000 are shown. As Figures 5.13 through 5.28 show the outputs of the ANN model followed the targets precisely. It shows that the resulting NN-based model can predict the

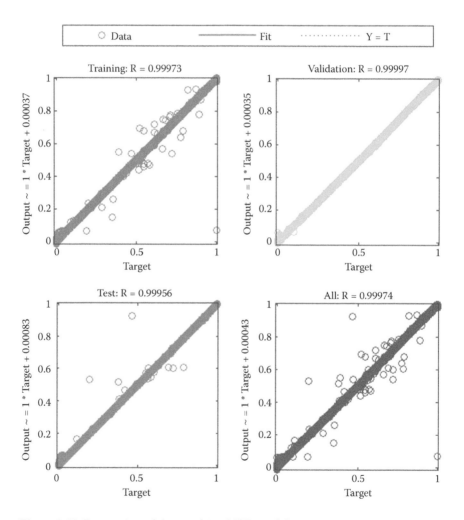

*Figure 5.12* Regression of the resulting MLP model.

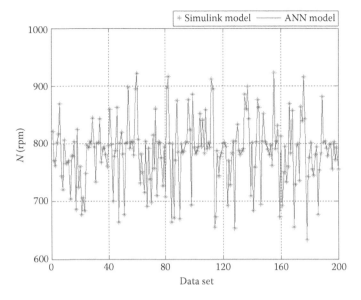

*Figure 5.13* Comparison between outputs of the Simulink® and ANN models for the rotational speed.

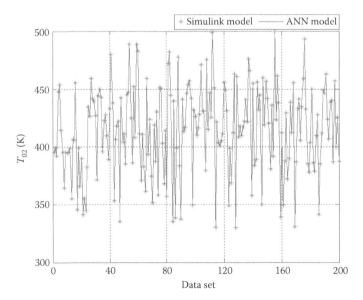

*Figure 5.14* Comparison between outputs of the Simulink® and ANN models for the compressor outlet temperature.

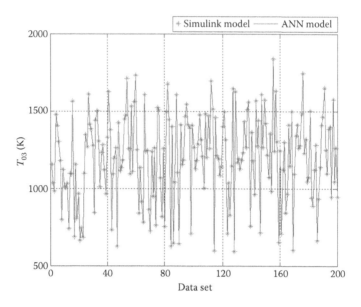

***Figure 5.15*** Comparison between outputs of the Simulink® and ANN models for the turbine inlet temperature.

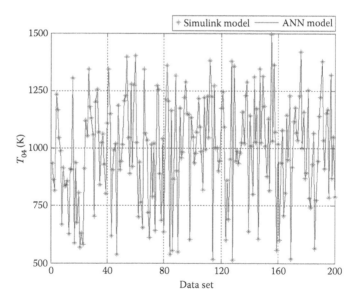

***Figure 5.16*** Comparison between outputs of the Simulink® and ANN models for the turbine outlet temperature.

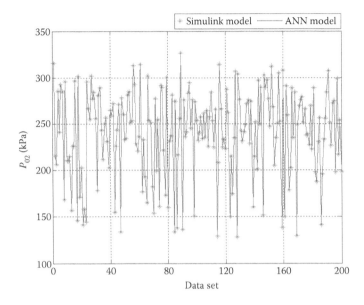

*Figure 5.17* Comparison between outputs of the Simulink® and ANN models for the compressor outlet pressure.

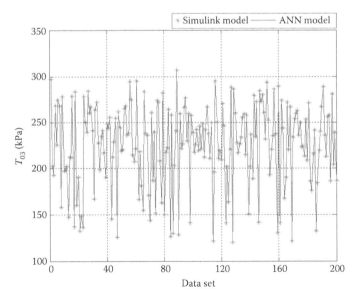

*Figure 5.18* Comparison between outputs of the Simulink® and ANN models for the turbine inlet pressure.

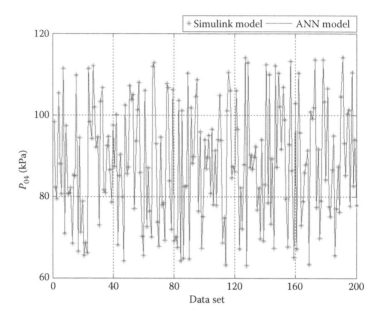

***Figure 5.19*** Comparison between outputs of the Simulink® and ANN models for the turbine outlet pressure.

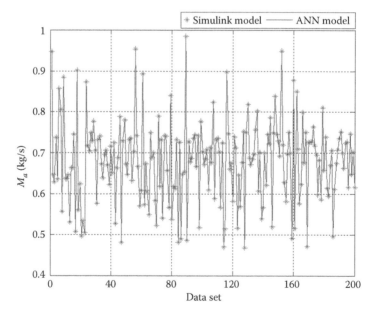

***Figure 5.20*** Comparison between outputs of the Simulink® and ANN models for the air mass flow rate.

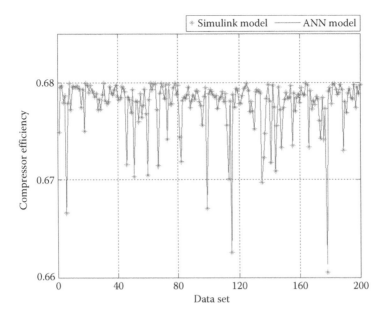

***Figure 5.21*** Comparison between outputs of the Simulink® and ANN models for the compressor efficiency.

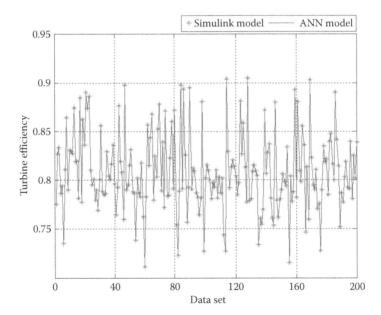

***Figure 5.22*** Comparison between outputs of the Simulink® and ANN models for the turbine efficiency.

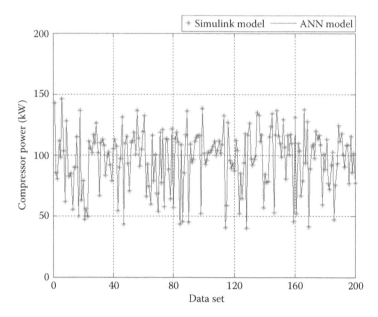

***Figure 5.23*** Comparison between outputs of the Simulink® and ANN models for the compressor power.

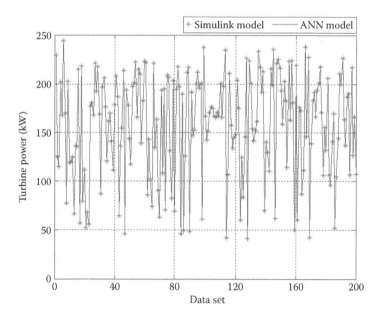

***Figure 5.24*** Comparison between outputs of the Simulink® and ANN models for the turbine power.

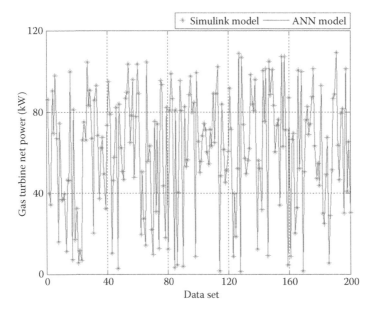

***Figure 5.25*** Comparison between outputs of the Simulink® and ANN models for the GT net power.

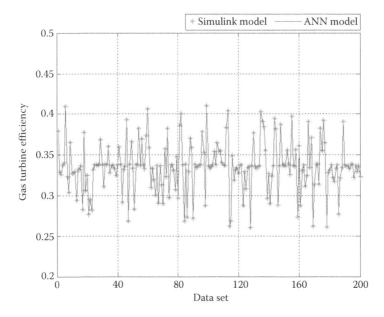

***Figure 5.26*** Comparison between outputs of the Simulink® and ANN models for the GT efficiency.

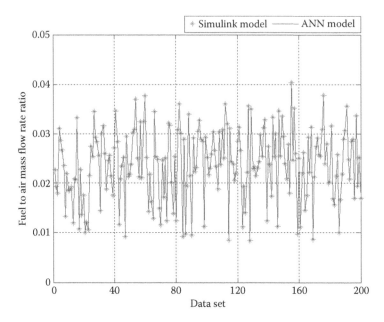

***Figure 5.27*** Comparison between outputs of the Simulink® and ANN models for the fuel to air mass flow rate ratio.

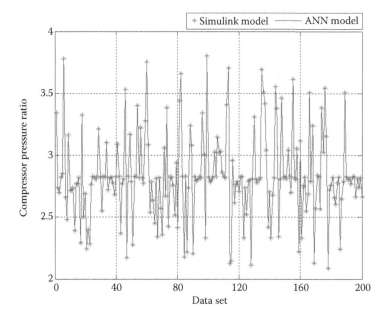

***Figure 5.28*** Comparison between outputs of the Simulink® and ANN models for the compressor pressure ratio.

reaction of the system to changes in input parameters with high accuracy and is capable of system identification with high reliability.

## 5.4 Summary

ANN has been used as a robust and reliable technique for system identification and modeling of complex systems with nonlinear dynamics such as GTs. It can provide feasible solutions to the problems that cannot be solved by conventional mathematical methods. ANN-based techniques can be applied to the systems through a variety of approaches that include different structures and training methods.

In this chapter, first a Simulink model of a low-power GT was developed based on thermodynamic and energy balance equations. In the next step, a new ANN-based methodology was applied to offline system identification of the GT. A comprehensive computer program code was generated and run in MATLAB environment using the obtained data from the Simulink model. Code generation was on the basis of combinations of various training functions, number of neurons and type of transfer functions. The methodology provided a comprehensive view of the performance of over 18,720 ANN models for system identification of the single-shaft GT.

The resulting model showed that the ANN-based method can be applied reliably for system identification of GTs. It can precisely predict output parameters of the GT based on the changes in the inputs of the system. The methodology applied in this chapter, can also be used to predict performance of similar GT systems with high accuracy when training from real data obtained from this type of GT. This methodology is particularly useful when real data is only available over a limited operational range. It was also observed that *trainlm* has a superior performance in terms of minimum *MSE*, compared with each of the other training functions.

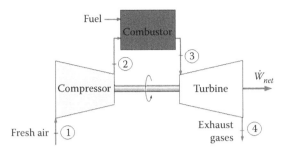

**Figure 1.1** Schematic of a typical single-shaft GT.

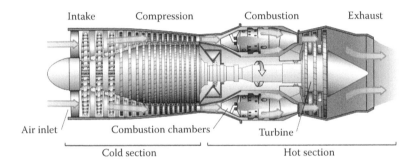

**Figure 1.3** A typical single-spool turbojet engine. (From Wikimedia Commons. 2012. [Online]. Available: http://commons.wikimedia.org.)

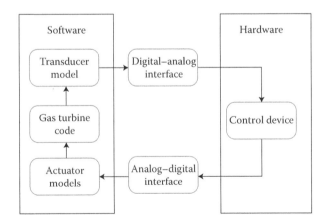

**Figure 2.2** The diagram for how real-time simulation software interacted with hardware control devices. (From S. M. Camporeale, B. Fortunato, and M. Mastrovito, *ASME Journal of Engineering for Gas Turbines and Power*, vol. 128, no. 3, pp. 506–517, 2006.)

**Figure 3.2** Block diagram of an ANN-based aero GT model for system optimization consists of minimizing fuel while maintaining thrust. (From N. Dodd and J. Martin, *Computing & Control Engineering Journal*, vol. 8, no. 3, pp. 129–135, 1997.)

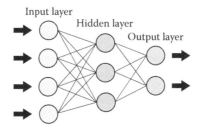

**Figure 4.1** A simple structure of an ANN with input, hidden, and output layers.

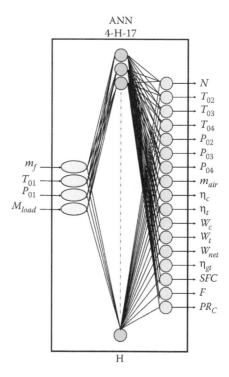

**Figure 5.9** A schematic of the ANN structure for the GT engine.

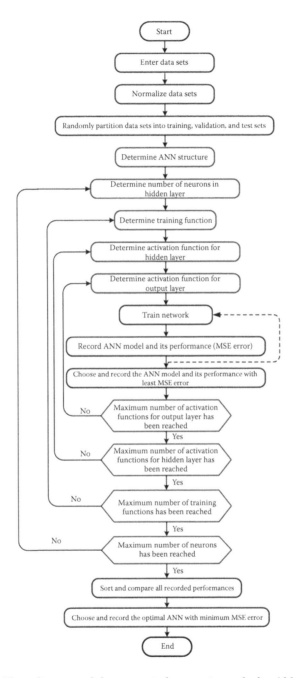

*Figure 5.10* Flow diagram of the generated computer code for ANN-based system identification of the GT.

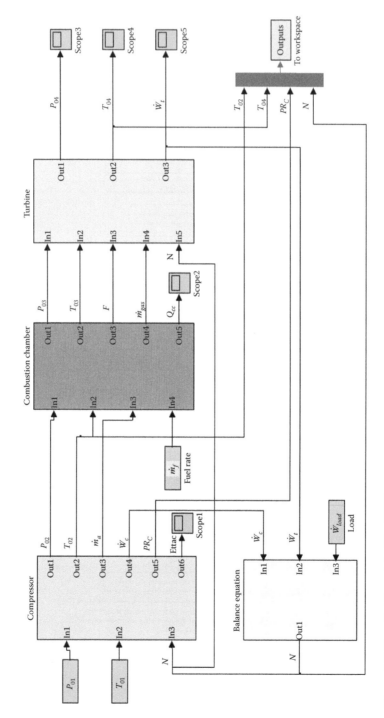

*Figure 6.7* Simulink® model of the IPGT.

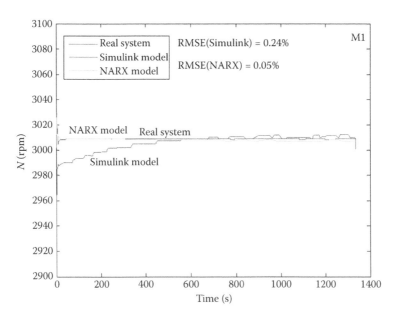

**Figure 6.11** Variations of rotational speed for the maneuver M1 for the real system, Simulink® model, and NARX model.

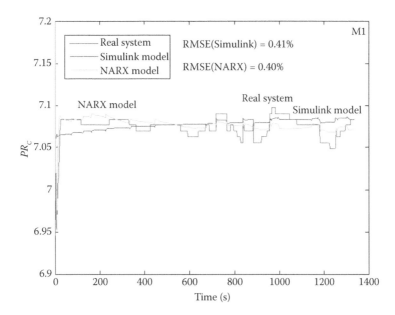

**Figure 6.12** Variations of compressor pressure ratio for the maneuver M1 for the real system, Simulink® model, and NARX model.

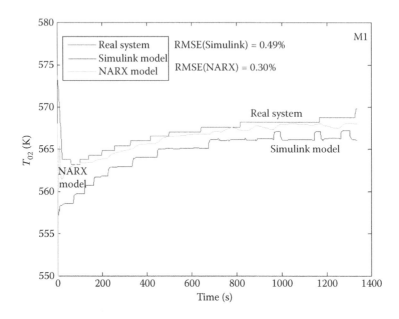

*Figure 6.13* Variations of compress outlet temperature for the maneuver M1 for the real system, Simulink® model, and NARX model.

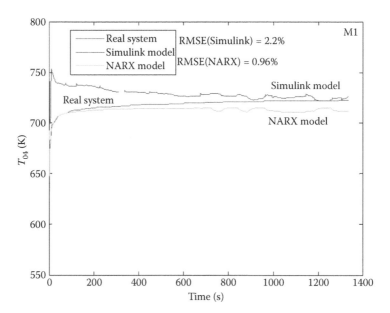

*Figure 6.14* Variations of turbine outlet temperature for the maneuver M1 for the real system, Simulink® model, and NARX model.

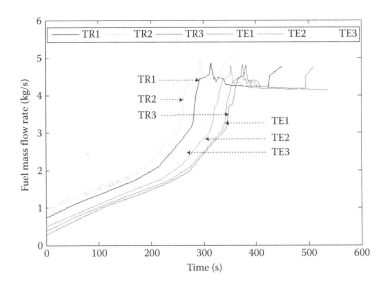

**Figure 7.1** Trend over time of mass flow rate.

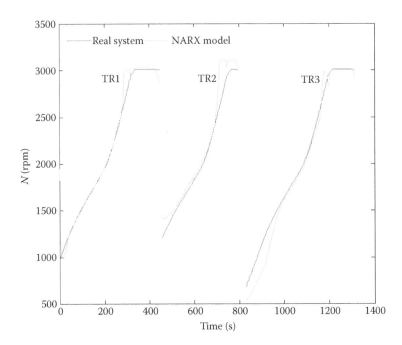

**Figure 7.4** Variations of rotational speed $N$ for the training maneuvers TR1, TR2, and TR3.

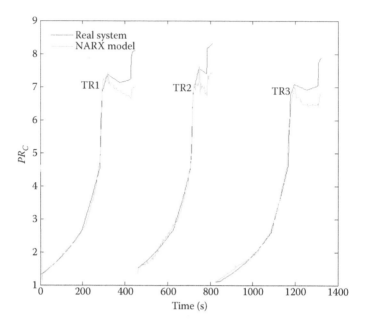

*Figure 7.5* Variations of compressor pressure ratio $PR_C$ for the training maneuvers TR1, TR2, and TR3.

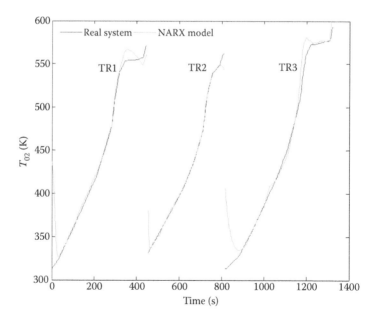

*Figure 7.6* Variations of compress outlet temperature $T_{02}$ for the training maneuvers TR1, TR2, and TR3.

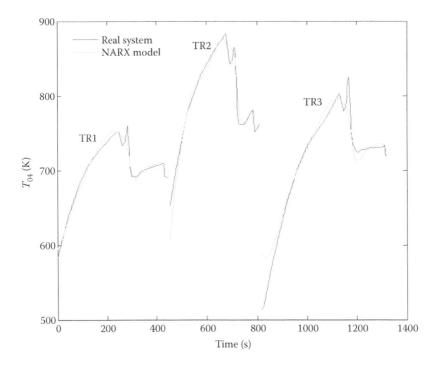

*Figure 7.7* Variations of turbine outlet temperature $T_{04}$ for the training maneuvers TR1, TR2, and TR3.

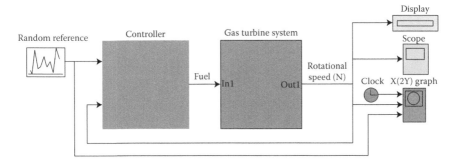

*Figure 8.1* The closed-loop diagram of the control system for the GT engine system.

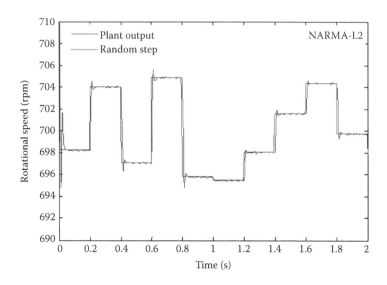

**Figure 8.18** Response of GT system with NARMA-L2 controller to random step inputs.

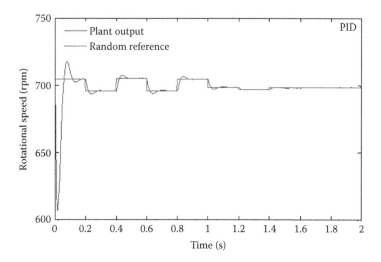

**Figure 8.22** Response of GT system with PID controller to random step inputs.

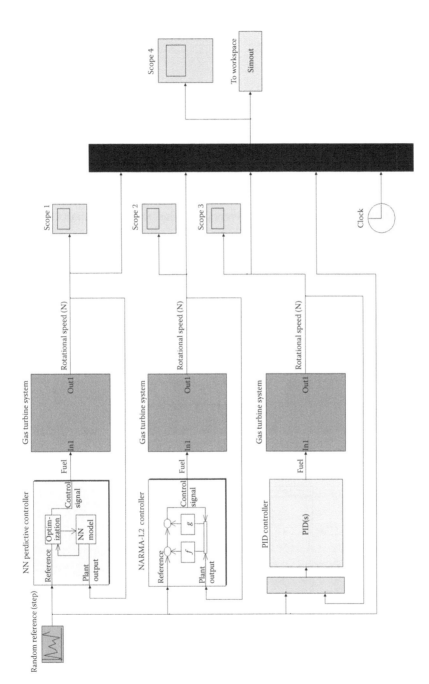

*Figure 8.23* Simulink® model of the ANN-based MPC, NARMA-L2, and PID controllers for a single-shaft GT.

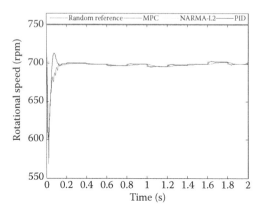

**Figure 8.24** Performances of three different controllers for a single-shaft GT.

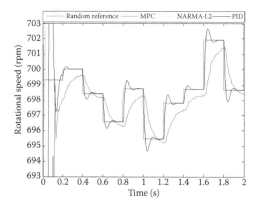

**Figure 8.25** A close-up perspective of the performances of three different GT controllers.

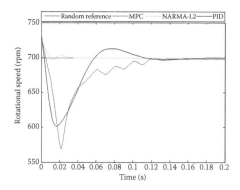

**Figure 8.26** A close-up perspective of the initial responses of three different GT controllers.

# chapter six

# *Modeling and simulation of dynamic behavior of an IPGT*

> The role of the infinitely small in nature is infinitely great.
>
> **Louis Pasteur**
> *French Chemist and Microbiologist, 1822–1895*

Modeling and simulation of industrial systems such as GTs is a significant methodology for system optimization. A GT model can be employed to clarify details of design strategies, manufacturing procedures, operating maneuvers, and even maintenance guidelines. Using black-box approach as a branch of artificial intelligence has opened a new horizon in the area of modeling and simulation of industrial systems. Black-box methodology is used to disclose the relationships between variables of the system using the measured operational data or data generated by means of a simulation tool. ANN, as a data-driven model, is one of the most significant methods in black-box modeling.

Majority of both, white-box and black-box models have been built based on the steady-state operation of GTs, when GTs have already passed the start-up procedure and run in a stable mode. Unfortunately, the available research sources lack enough investigation on modeling and simulation of GT transient behavior and start-up operation, especially for IPGTs.

Because of the importance of transient behavior of GTs during start-up and its direct effect on GT performance and lifetime, extensive research is still necessary to fill the existing information gaps. According to the available information, no model has been developed so far to simulate GT transient behavior during start-up and near full-speed operation by using Simulink® and NARX models. One of the few examples of such simulation models is documented by Asgari et al. [230], where an NARX model was set up and optimized for the simulation of the start-up operation of an IPGT. Moreover, another challenging issue is the use of field data for model development and testing. Therefore, the set-up and application of these models can help in understanding and analyzing the transient behavior of GTs.

In this chapter, two separate simulation models using both white-box and black-box methods are built to simulate very low-power operating region for an IPGT. The modeling and simulation are carried out on the basis of the experimental time-series data sets obtained from an IPGT located in Italy [231]. The specifications of the GT are described in Section 6.1. The subject of Section 6.2 is data acquisition and preparation. Sections 6.3 and 6.4 present the physics-based modeling approach in Simulink and the set-up of a black-box model by using NARX modeling approach, respectively. The comparison of all the significant measured and predicted variables and a summary of the results are presented in Sections 6.5 and 6.6, respectively.

## 6.1   GT specifications

The GT modeled in this chapter is the General Electric PG 9351FA, which is a heavy-duty single-shaft GT used for power generation. The main specifications of this IPGT are summarized in Table 6.1.

## 6.2   Data acquisition and preparation

The data sets used for model set-up and verification were taken experimentally during several start-up maneuvers. The data sets cover the range 420–3000 rpm. Power is also very low (less than 24 MW) compared to the nominal power equal to 260 MW, approximately. Therefore, these data are representative of the operating conditions during start-up and also account for all the conditions related to this type of transient operation (e.g., bleed valve opening, IGV control, etc.). The amount of time elapsed since last shutdown greatly influences startup time. It takes longer to start GTs from cold conditions than from hot conditions. The definition of hot conditions varies by GT manufacturers; in general, the start-up conditions can be categorized as

- Hot start-up: The GT was shut down for just a short time before start-up (generally within 8–16 h of GT shutdown).
- Warm start-up: The GT was shut down for a longer time compared to hot start-up conditions, but less than 1 day of GT shutdown).
- Cold start-up: The GT was shut down some days before start-up.

*Table 6.1* GT specifications

| GT type | Number of shafts | Rotational speed (rpm) | Pressure ratio | TIT (°C) | TOT (°C) | Airflow rate (kg/s) | Power (kW) | HR (kJ/ kWh) | Efficiency % |
|---------|------------------|------------------------|----------------|----------|----------|---------------------|------------|--------------|--------------|
| GE 9351FA | 1 | 3000 | 15.8 | 1327 | 599 | 648 | 259500 | 9643 | 37.3 |

The data sets used in this chapter refer to cold start-up. Moreover, each of the data sets may fall into different combinations of the following conditions:

- If the starter is on or off: 1 or 0
- If the GT is connected to the grid or not: 1 or 0
- If customer trip happens or not: 1 or 0
- If the flame is on or off: 1 or 0

To model the GT, the operating conditions sketched below are considered:

- The starter is off: 0
- The GT is connected/disconnected to/from the grid: 1 or 0
- Customer trip does not happen: 0
- The flame is on: 1

For instance, the maneuver [0 1 0 1] refers to the situation, when the starter is off, the GT is connected to the grid, customer trip does not happen, and the flame is on.

The measured time-series data sets [0 1 0 1] and [0 0 0 1] are called M1 and M2, respectively, and are used for Simulink model tuning. The two other data sets including [0 1 0 1 a] and [0 1 0 1 b] which were used for verification of the models are specified as M3 and M4. Table 6.2 shows more details of these data and the operational range for the input parameters. The time step for the data acquisition is 1 s. Figure 6.1 shows the variations of load for other maneuvers. With respect to M2 and M4, the changes of the load of M3 are smaller and as Table 6.2 indicates M4 is the longest maneuver. Load for M1 is very low and nearly constant.

*Table 6.2* Time-series data sets for different maneuvers

| Maneuver | Type of data set | Number of data | Operational range of the inputs | | | |
|---|---|---|---|---|---|---|
| | | | $T_{01}$ (K) | $P_{01}$ (Pa) | $\dot{m}_f$ (kg/s) | $\dot{W}_{load}$ (MW) |
| M1 | [0 0 0 1] | 1336 | [296.48; 301.50] | [99570; 99909] | [3.74; 4.60] | $\approx$0.3 |
| M2 | [0 1 0 1] | 1165 | [297.04; 303.15] | [99570; 99670] | [3.99; 4.50] | [3.30; 18.70] |
| M3 | [0 1 0 1 a] | 1506 | [290.37; 295.37] | [101940; 102040] | [4.42; 4.80] | [4.42; 4.76] |
| M4 | [0 1 0 1 b] | 3296 | [299.26; 302.59] | [100950; 101070] | [4.27; 6.90] | [18.20; 23.70] |

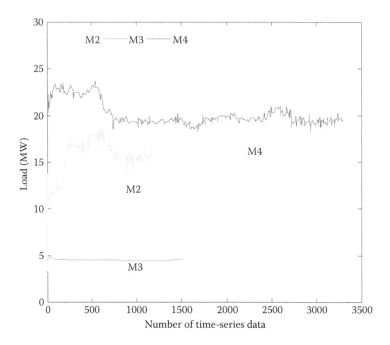

**Figure 6.1** Variations of load for different maneuvers.

## 6.3   Physics-based model of IPGT by using Simulink®: MATLAB®

The Simulink model of the GT was built by using the operational time-series data sets and by employing the thermodynamic equations for its components. The data sets were used to approximate the correlations between corrected parameters in the compressor and turbine. The approximations were obtained by using the Curve Fitting Tool in MATLAB®. The set-up of the Simulink model consisted of modeling the four main components including compressor, combustion chamber, turbine, and rotational parts dynamics. Each component was considered as a single block, stacking dynamic behavior of all the individual stages into a single block with only the inlet and exit conditions of the component. The main thermodynamic equations used in the physics-based model and the modeling assumptions and the parameter values are discussed in the following:

- Compressor

$$T_{02} = T_{01} + \frac{T_{01}}{\eta_c}\left[\left(\frac{P_{02}}{P_{01}}\right)^{(\gamma_{air}-1)/\gamma_{air}} - 1\right] \qquad (6.1)$$

$$\gamma_{air} = \frac{C_{p\,air}}{C_{v\,air}} \tag{6.2}$$

$$\dot{W}_c = \dot{m}_{air} C_{p\,air} (T_{02} - T_{01}) \tag{6.3}$$

- Turbine

$$T_{04} = T_{03} - T_{03}\eta_t \left[ 1 - \left( \frac{P_{04}}{P_{03}} \right)^{(\gamma_{gas}-1)/\gamma_{gas}} \right] \tag{6.4}$$

$$\gamma_{gas} = \frac{C_{p\,gas}}{C_{v\,gas}} \tag{6.5}$$

$$\dot{W}_t = \dot{m}_{gas} (C_{p\,gas\_03} T_{03} - C_{p\,gas\_04} T_{04}) \tag{6.6}$$

- Combustor

$$\frac{1}{F} = \dot{m}_a / \dot{m}_f = \frac{\eta_{cc} * LHV}{(C_{p\,gas\_03} T_{03} - C_{p\,air\_02} T_{02})} - 1 \tag{6.7}$$

$$P_{03} = P_{02}(1 - \xi_{cc}) \tag{6.8}$$

- Equilibrium (balance) equation

$$(\dot{W}_t - \dot{W}_c - \dot{W}_{load}) = (2\pi/60)^2 \, I \, N \, (dN/dt) \tag{6.9}$$

### 6.3.1   Measured parameters

The parameters which were measured directly from the IPGT include rotational speed ($N$), alternator power ($\dot{W}_{load}$), ambient pressure ($P_{00}$), ambient temperature ($T_{00}$), compressor inlet stagnation pressure ($P_{01}$), compressor inlet temperature ($T_{01}$), compressor outlet stagnation pressure ($P_{02}$), compressor outlet temperature ($T_{02}$), turbine outlet temperature ($T_{04}$), and fuel flow rate ($\dot{m}_f$).

### 6.3.2   Calculated or estimated parameters

To formulate the correlation between corrected parameters of the compressor and turbine components of the GT, to be used in the Simulink models, calculation, or estimation of some unmeasured parameters was unavoidable. These parameters can be extracted by employing thermodynamic relationships or general experimental results about GTs.

### 6.3.2.1 Turbine inlet stagnation pressure

As Equation 6.8 shows, in practical applications, $P_{03}$ can be approximated by considering a linear decrease with respect to $P_{02}$. The loss pressure in the combustion chamber ($\xi_{cc}$) is about 3%.

### 6.3.2.2 Turbine outlet stagnation pressure

As a practical routine, turbine outlet stagnation pressure is estimated according to Equation 6.10.

$$P_{04} \simeq P_{00} + C \tag{6.10}$$

The coefficient $C$ is assumed a constant value and approximated as $C = 0.03$ bar.

### 6.3.2.3 Specific heat of air and gas at constant pressure

Specific heat of air and gas at constant pressure is calculated based on the fact that it is a function of average temperatures during the compression and expansion processes in the compressor and turbine. After the determination of $C_p$, specific heat of air and gas at constant volume is calculated using the following equations:

$$\gamma = \frac{C_p}{C_v} \tag{6.11}$$

$$R = C_p - C_v \tag{6.12}$$

$C_p$ is calculated using Equations 6.13 through 6.16. $T_a$ and $T_g$ refer to the average temperatures during the compression and expansion processes in the compressor and turbine, respectively [47].

If $T_a < 800K$

$$C_{pair} = 1018.9 - 0.13784*T_a + 1.9843E - 04*T_a^2 + 4.2399E$$
$$- 07*T_a^3 - 3.7632E - 10*T_a^4 \tag{6.13}$$

$$C_{pgas} = C_{pair} + \left(\frac{F}{1+F}\right)*(-359.494 + 4.5164*T_g + 2.8116E - 03*T_g^2$$
$$- 2.1709E - 05*T_g^3 + 2.8689E - 08*T_g^4 - 1.2263E - 11*T_g^5) \tag{6.14}$$

If $T_a > 800K$

$$C_{p_{air}} = 798.65 + 0.5339*T_a - 2.2882E - 04*T_a^2 + 3.7421E - 08*T_a^3 \quad (6.15)$$

$$C_{p_{gas}} = C_{p_{air}} + \left(\frac{F}{1+F}\right)*(1088.8 - 0.1416*T_g + 1.916E - 03*T_g^2$$
$$- 1.2401E - 06*T_g^3 + 3.0669E - 10*T_g^4 - 2.6117E - 14*T_g^5) \quad (6.16)$$

### 6.3.2.4    Turbine inlet temperature and mass flow rate of air

When $\dot{W}_{load}$ is approximately constant (i.e., acceleration is zero), Equation 6.9 can be written as:

$$\dot{W}_t = \dot{W}_c + \dot{W}_{load} \quad (6.17)$$

Then, by replacing $\dot{W}_c$ from Equation 6.3, $\dot{W}_t$ can be determined, and Equation 6.6 can be written as follows:

$$T_{03} = \frac{C_{p_{gas\_04}}}{C_{p_{gas\_03}}} T_{04} + \frac{\dot{W}_t}{\dot{m}_{gas} C_{p_{gas\_03}}} \quad (6.18)$$

in which

$$\dot{m}_{gas} = \dot{m}_f + \dot{m}_a \quad (6.19)$$

Besides, Equation 6.7 can be written as

$$T_{03} = \frac{C_{p_{air\_02}}}{C_{p_{gas\_03}}} T_{02} + \frac{\eta_{cc}*LHV}{C_{p_{gas\_03}}*(1 + \dot{m}_a/\dot{m}_f)} \quad (6.20)$$

Finally, Equations 6.18 and 6.20 can be solved for $T_{03}$ and $\dot{m}_a$. For this purpose, a computer code was written and run in MATLAB, and these parameters were calculated for the available data sets.

### 6.3.2.5    Efficiency and corrected parameters
### of the compressor and turbine

The code already written for calculation of $T_{03}$ and $m_a$ was developed to calculate efficiency and corrected parameters for both the compressor and the turbine. Compressor efficiency ($\eta_c$) and turbine efficiency ($\eta_T$) are calculated using the following equations:

$$\eta_c = T_{01} * \left[ \left( \left( \frac{P_{02}}{P_{01}} \right) \right)^{(\gamma_{air} - 1)/\gamma_{air}} - 1 \right] / (T_{02} - T_{01}) \qquad (6.21)$$

$$\eta_T = (T_{03} - T_{04}) / \left[ T_{03} * \left( 1 - \left( \frac{P_{04}}{P_{03}} \right)^{(\gamma_{gas} - 1)/\gamma_{gas}} \right) \right] \qquad (6.22)$$

### 6.3.3   Model architecture

Figure 6.2 shows the block diagram of the GT system. It includes four inputs and four outputs. The inputs are compressor inlet temperature ($T_{01}$), compressor inlet stagnation pressure ($P_{01}$), fuel flow rate ($\dot{m}_f$), and network load ($\dot{W}_{load}$). The outputs consist of rotational speed ($N$), compressor pressure ratio ($PR_C$), compressor outlet temperature ($T_{02}$), and turbine outlet temperature ($T_{04}$).

Figure 6.2 outlines a summarized block diagram of the Simulink model in a MATLAB environment including inputs and outputs, while Figures 6.3 through 6.6 show subsystems of the model. Figure 6.7 indicates a block diagram of the model with more details to highlight the information flow.

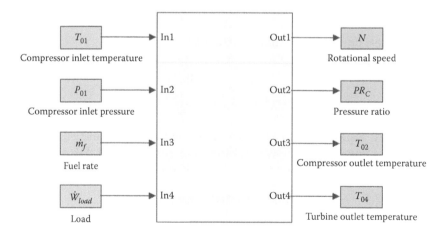

*Figure 6.2* Block diagram of the Simulink® model of the IPGT.

### 6.3.4 Discussion on physics-based modeling approach

As outlined in the previous equations in this chapter, only one equation is used to account for system dynamics, that is, torque balance, since this is usually recognized as the most influencing factor on transient behavior. All other equations represent steady-state correlations among the different thermodynamic quantities, calculated by using the performance maps obtained by fitting the experimental data used for model tuning through the Curve Fitting Tool in MATLAB. This is a key step in tuning the model, which allows reproducing actual GT behavior in the operating region under consideration. Fitting the experimental data also allows smoothing measurement uncertainty, which affects the measured data sets taken by using standard GT sensors. Another innovative aspect of the developed model is the iterative procedure adopted for estimating turbine inlet temperature and inlet mass flow rate, as discussed in Section 6.3.2.4. This procedure assures that, at any time point, the energy balance is satisfied

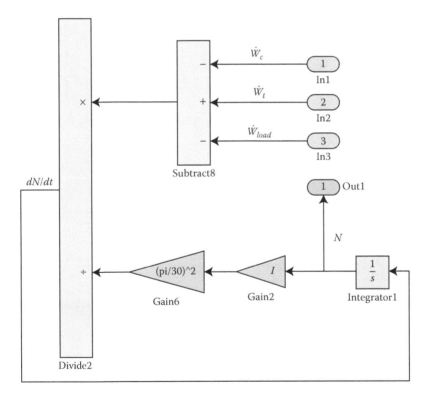

*Figure 6.3* Subsystem number 1 of the IPGT Simulink® model for modeling the balance equation.

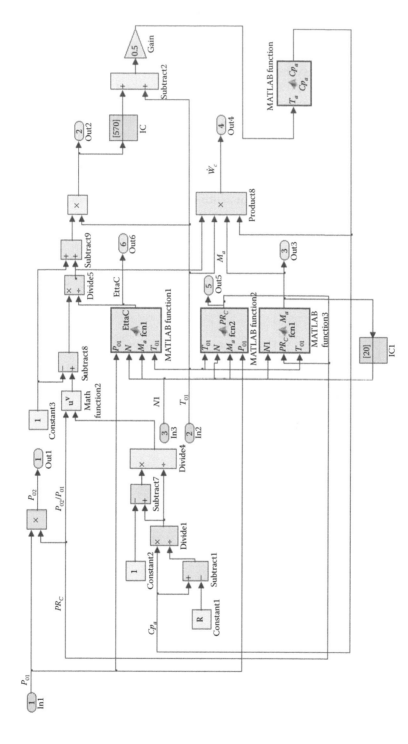

***Figure 6.4*** Subsystem number 2 of the IPGT Simulink® model for creating the compressor model.

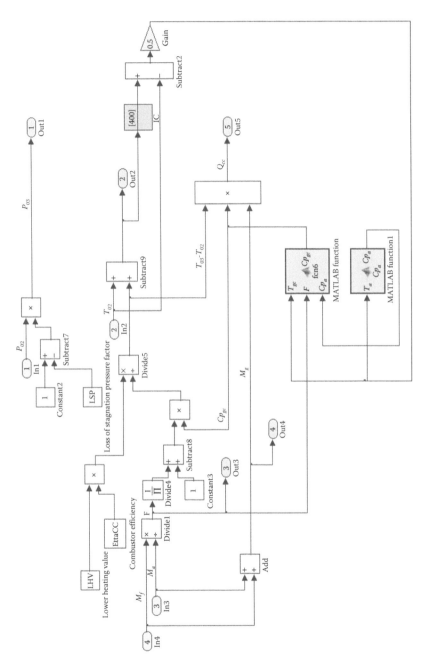

*Figure 6.5* Subsystem number 3 of the IPGT Simulink® model for creating the combustion chamber model.

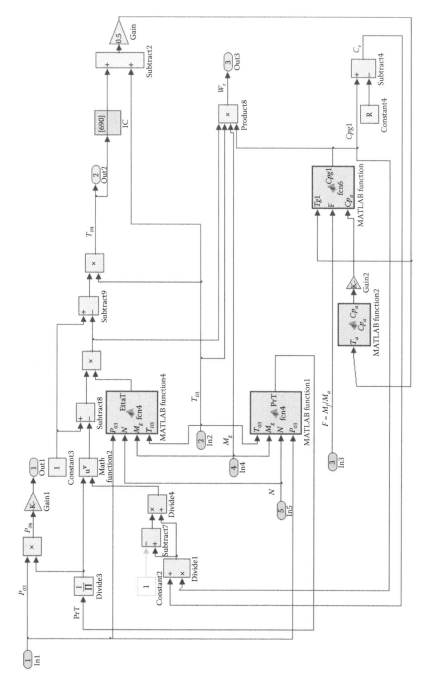

***Figure 6.6*** Subsystem number 4 of the IPGT Simulink® model for creating the turbine model.

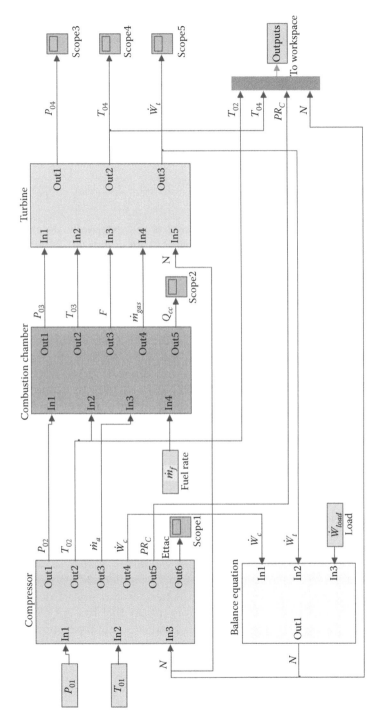

*Figure 6.7* **(See color insert.)** Simulink® model of the IPGT.

and in agreement with the instant measured values. Finally, it is stressed that this dynamic model has been developed and it will be validated in the next sections by using field data.

## 6.4   NARX model of IPGT

NARX is a recurrent dynamic network commonly used in time-series modeling. It includes feedback connections enclosing several layers of the network. The defining equation for the NARX model is as follows [212]:

$$y(t) = f(y(t-1),\ y(t-2),\ \dots, y(t-n_y), u(t-1),\ u(t-2),\ \dots,\ u(t-n_u)) \quad (6.23)$$

where, $y$ is the variable of interest, and u is the externally determined variable. The next value of the dependent output signal $y(t)$ is regressed on previous values of the output signal and previous values of an independent (exogenous) input signal. NARX models can be implemented by using an FFNN to approximate the function $f$ [212]. NARX networks have many applications. For instance, they can be used for nonlinear filtering of noisy input signals or prediction of the next value of the input signal. However, the most significant application of NARX network is to model nonlinear dynamic systems [212].

The NNT in MATLAB was employed to tune the NARX models by using measured time-series data sets. An NARX model was trained separately for M1 to predict outputs for [0 0 0 1] start-up condition. For the whole time-series data sets related to M2, M3, and M4 maneuvers, another NARX model was trained and the final NARX models were obtained after trial-and-error efforts for getting reliable and accurate models in terms of accuracy of the trends and RMSE for output parameters. The resulting model was tested against each of the M2, M3, and M4 maneuvers separately. Inputs and outputs of the NARX models are the same corresponding parameters as in the Simulink model.

Figures 6.8 and 6.9 show the block diagram and the closed-loop structure of the NARX model in MATLAB environment. As it can be seen from Figure 6.9, the best result for both NARX models is related to networks with one hidden layer of nine neurons, using Levenberg–Marquardt backpropagation (*trainlm*) as the training function, and a tapped delay line with delays from 1 to 2 at the input. This means that the NARX model makes use of the regressed outputs $y(t-1)$ and $y(t-2)$ at time points $(t-1)$ and $(t-2)$.

Figure 6.10 shows the details of the final trained network. Performance of the NARX for training, validation, and test is also shown in the figure. As it can be seen, 13 iterations were required so that the validation performance error reached the minimum. MSE of the performance at this point

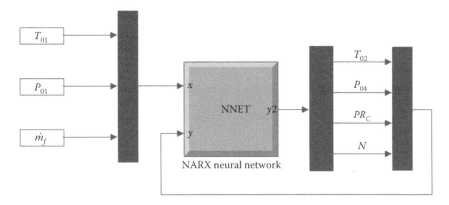

**Figure 6.8** Block diagram of complete NARX model of the IPGT.

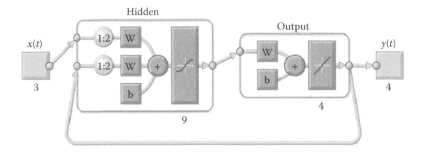

**Figure 6.9** NARX model of the IPGT.

was quite low (<0.014). The training had continued for six more iterations before the training stopped. Values of the measured data and the predictions of NARX models were compared on the basis of RMSE already defined according to Equation 4.3 in Chapter 4.

## 6.5    Comparison of physics-based and NARX models

The trend over time of the prediction of the two simulation models (physics-based model developed in Simulink and NARX model) for the four outputs (rotational speed, pressure ratio, compressor outlet temperature, and turbine outlet temperature) is compared to the trend over time of measured data. The comparison is made both for the "training" curves M1 and M2 (Figures 6.11 through 6.14 and 6.15 through 6.18, respectively) and for the curves M3 and M4 used to assess the generalization capability of the simulation models (Curve M3 in Figures 6.19 through 6.22

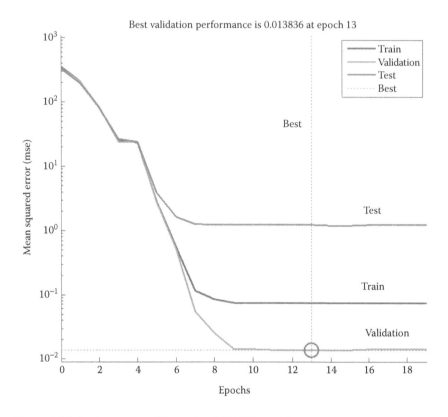

*Figure 6.10* Performance of the trained NARX model.

and curve M4 in Figures 6.23 through 6.26). Figure 6.27 summarizes the results in terms of RMSE for all the maneuvers, to allow a synoptic view. It should be noted that, during the simulation phase, the NARX model is fed with the regressed outputs at time points $y(t-1)$ and $y(t-2)$ estimated by the NARX model itself at antecedent time steps.

As it can be seen, M1 is reproduced very accurately by both models. The maximum RMSE of Simulink and NARX models are 2.2% and 0.96% (both for $T_{04}$) respectively. M2 is also reproduced with a satisfactory prediction. The maximum values of RMSE of Simulink and NARX models for this maneuver are 4.3% and 2.1% (both for $PR_C$). M3 is also simulated with acceptable accuracy. The maximum *RMSE* of Simulink and NARX models are 3.9% and 2.8% (both for $PR_C$) respectively. The results for M4 are also satisfactory enough for prediction of GT dynamics. The maximum errors of Simulink and NARX models for this maneuver are 4% and 1.7%, both for $PR_C$. It can be noticed that despite higher errors at the beginning of the simulation until the stabilization of the response, the RMSE was satisfactory in the Simulink model. The *RMSE* of rotational

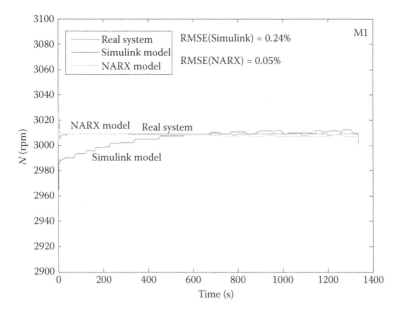

*Figure 6.11* (**See color insert.**) Variations of rotational speed for the maneuver M1 for the real system, Simulink® model, and NARX model.

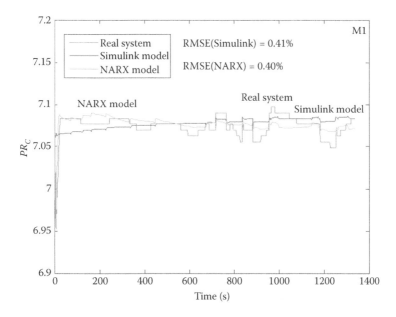

*Figure 6.12* (**See color insert.**) Variations of compressor pressure ratio for the maneuver M1 for the real system, Simulink® model, and NARX model.

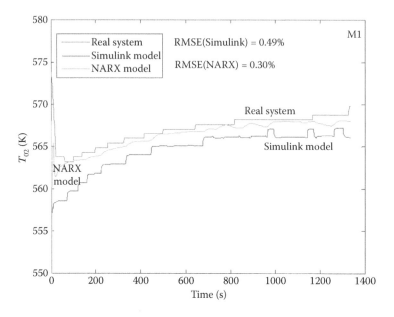

*Figure 6.13* **(See color insert.)** Variations of compress outlet temperature for the maneuver M1 for the real system, Simulink® model, and NARX model.

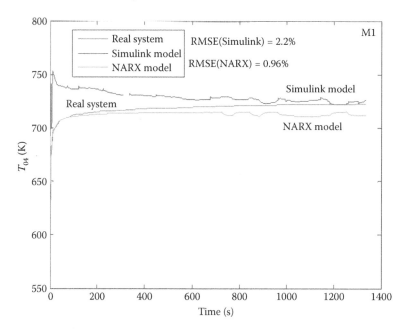

*Figure 6.14* **(See color insert.)** Variations of turbine outlet temperature for the maneuver M1 for the real system, Simulink® model, and NARX model.

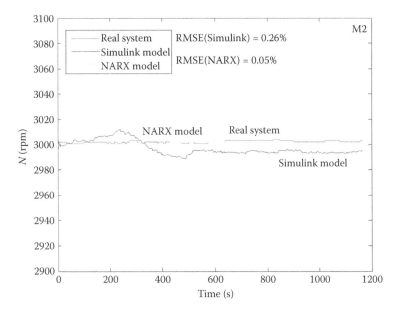

**Figure 6.15** Variations of rotational speed for the maneuver M2 for the real system, Simulink® model, and NARX model.

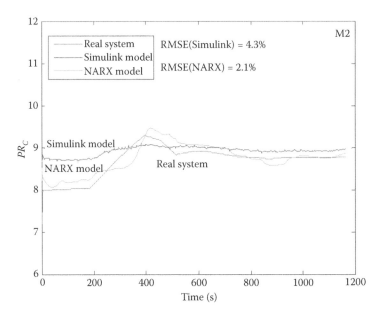

**Figure 6.16** Variations of compressor pressure ratio for the maneuver M2 for the real system, Simulink® model, and NARX model.

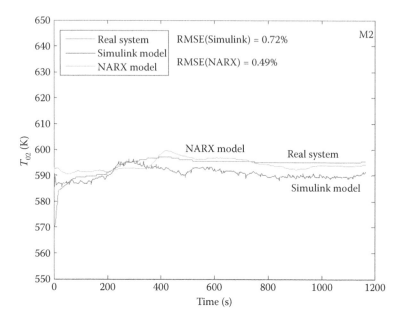

**Figure 6.17** Variations of compress outlet temperature for the maneuver M2 for the real system, Simulink® model, and NARX model.

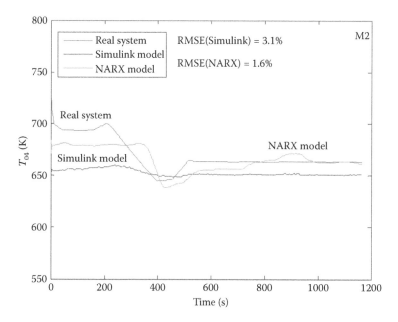

**Figure 6.18** Variations of turbine outlet temperature for the maneuver M2 for the real system, Simulink® model, and NARX model.

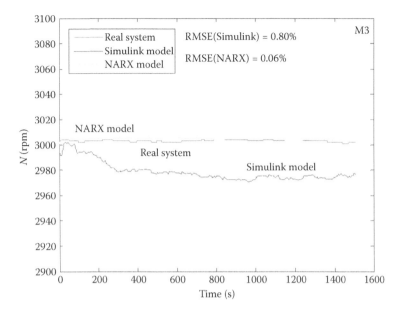

***Figure 6.19*** Variations of rotational speed for the maneuver M3 for the real system, Simulink® model, and NARX model.

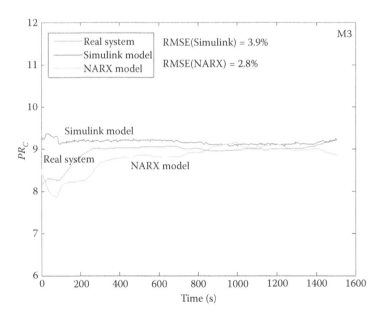

***Figure 6.20*** Variations of compressor pressure ratio for the maneuver M3 for the real system, Simulink® model, and NARX model.

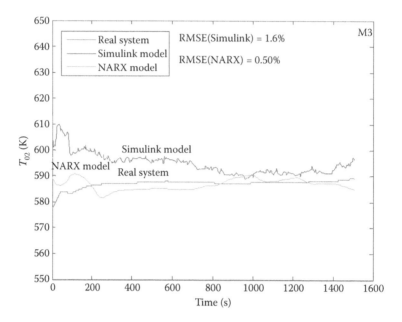

**Figure 6.21** Variations of compress outlet temperature for the maneuver M3 for the real system, Simulink® model, and NARX model.

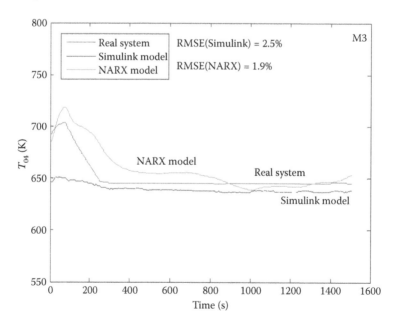

**Figure 6.22** Variations of turbine outlet temperature for the maneuver M3 for the real system, Simulink® model, and NARX model.

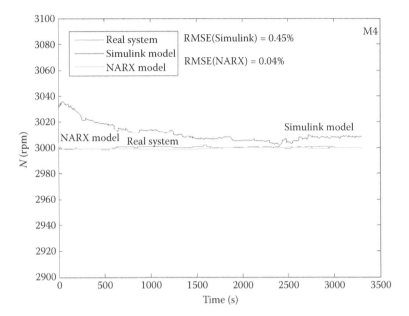

**Figure 6.23** Variations of rotational speed for the maneuver M4 for the real system, Simulink® model, and NARX model.

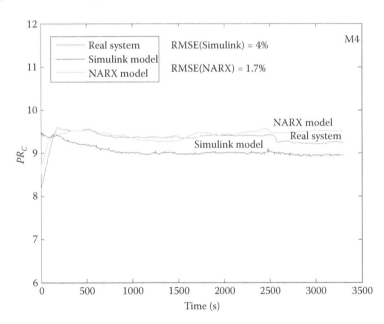

**Figure 6.24** Variations of compressor pressure ratio for the maneuver M4 for the real system, Simulink® model, and NARX model.

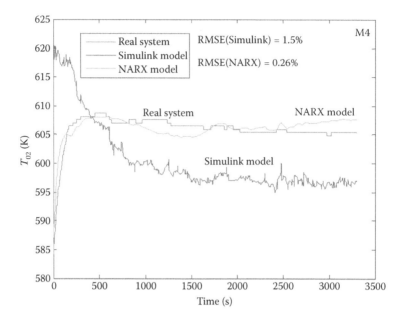

***Figure 6.25*** Variations of compress outlet temperature for the maneuver M4 for the real system, Simulink® model, and NARX model.

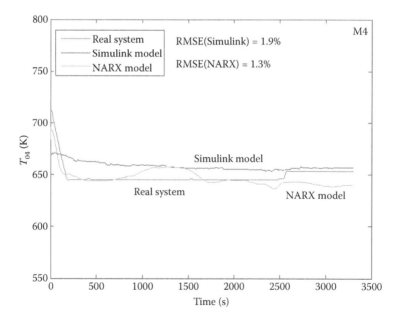

***Figure 6.26*** Variations of turbine outlet temperature for the maneuver M4 for the real system, Simulink® model, and NARX model.

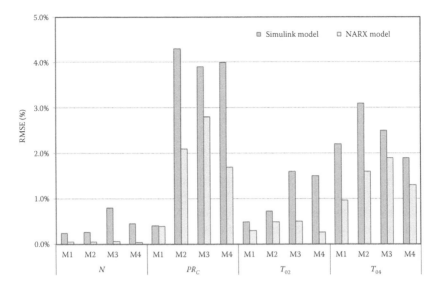

*Figure 6.27* RMSE (%) of the Simulink® and NARX models for main selected outputs of all the maneuvers.

speed, pressure ratio, compressor outlet temperature, and turbine outlet temperature for all maneuvers were equal to or less than 0.8%, 4%, 1.6%, and 3.1% respectively. The maximum error of the Simulink and NARX models were respectively 4.3% and 2.8%. Overall, the results show that both Simulink and NARX models can simulate and predict the dynamic behavior of the GT with acceptable accuracy. However, the NARX model showed higher accuracy compared to the Simulink model.

## 6.6    Summary

In this chapter, Simulink and NARX models of a heavy-duty single-shaft power plant GT were set up for simulating two different transient maneuvers in the very low-power operating region. The main objective was to explore and predict transient behavior of IPGTs. Thermodynamic and energy balance equations were employed to model the GT in Simulink-MATLAB environment. Correlations between corrected parameters of the GT components (compressor and turbine) were investigated by using measured data sets and by using Curve Fitting Tool in MATLAB. The same time-series data sets were employed to build NARX model for the IPGT. To verify the models, the resulting Simulink and NARX models were tested against two other time-series data sets. For this purpose, four important outputs from the IPGT models and their corresponding values from the measured data sets were compared, and the related results were captured and figured. These outputs were rotational speed, compressor pressure ratio, compressor outlet temperature, and turbine outlet temperature. The results showed that both Simulink and NARX models successfully captured dynamics of the system. They provided satisfactory prediction of the dynamic behavior of the GT for the studied maneuvers.

The simplicity of the thermodynamic model developed above is one of the strong points of this methodology. In fact, accurate modeling would have required the knowledge of bleed flows and IGV control. Since the information about bleed flows is usually unknown (they are confidential manufacturer's data) and IGV control logic is not always known (in this case, it was not available, since this information is confidential manufacturer's data as well), this lack of information was overcome by implicitly accounting for these effects (which mainly affect the inlet mass flow rate) by means of two innovative procedures:

- The performance maps, which relate the corrected parameters were obtained directly from measured data (only from the "training" data sets M1 and M2, and not from the "verification" data sets M3, and M4) by using the Curve Fitting Tool available in MATLAB. The fine-tuning of these correlations represented a key and challenging phase of the thermodynamic model set-up.
- The inlet mass flow rate and the turbine inlet temperature were estimated at each time step by means of an iterative procedure in Equations 6.18 through 6.20. A specific MATLAB routine was written and dedicated to this calculation.

The choice of developing a NN model to cover this range of operation goes in the same direction, that is, developing a simple model to reproduce a very complicated and usually difficult-to-model unsteady behavior. The results indicated that NARX approach modeled GT behavior with higher accuracy compared to Simulink approach. It was shown that ANN could be considered as a reliable and powerful tool for identification of systems dynamics. Moreover, the NARX approach to transient analysis may have the potential to provide some diagnostic information for the entire GT.

*chapter seven*

# Modeling and simulation of the start-up operation of an IPGT by using NARX models

The important thing is not to stop questioning.
Curiosity has its own reason for existing.

**Albert Einstein**
*German-American Physicist, 1879–1955*

Accurate modeling requires the knowledge of bleed flows and inlet guide vanes control. However, the operational modes with modern dry low nitrogen oxide (DLN) and dry low emission (DLE) systems also involving fuel splits and bleed action are very complex. Since such pieces of information are usually unknown, or they are confidential manufacturer's data, the adoption of a black-box approach allows the implicit incorporation of all these phenomena in a simple simulation model. For this reason, NARX models of start-up procedure for a heavy-duty IPGT are constructed in this chapter. The modeling and simulation are carried out on the basis of the experimental time-series data sets.

This chapter represents one of the few attempts to develop a dynamic model of the IPGT (and, in particular, for the start-up maneuver) by means of NARX models and validate it against experimental data taken during normal operation by means of standard measurement sensors and acquisition system. Building the required models in this specific area can be very effective in understanding and analyzing GT dynamics, and can also provide information about fault diagnostics.

The GT modeled in this chapter is the same IPGT as described in Table 6.1 of Chapter 6. It is a heavy-duty single-shaft GT for power generation (General Electric PG 9351FA). GT start-up procedure is described first, then the main steps for data acquisition and preparation, NARX modeling, and the results of the comparison of NARX prediction to experimental measurements are discussed. This chapter ends up with the summary and concluding remarks.

## 7.1 GT start-up

Start-up period is the operating period before the GT reaches stable combustion conditions. To start to work, GTs need an external source, such as an electrical motor or a diesel engine. GTs use a starter until the engine speed reaches a specific percentage of the design speed. Then, the engine can sustain itself without the power of the starter.

GT start-up procedure can be divided into four phases including dry cranking, purging, light-off, and acceleration to idle [232,233]. In dry cranking phase, the engine shaft is rotated by the starting system without any fuel feeding. In purging phase, residual fuel from the previous operation or failed start attempts is purged out of the fuel system. In this phase, the rotating speed is kept constant at a value, which ensures a proper mass flow rate through the combustion chamber, the turbine, and the heat recovery steam generator. During light-off, fuel is fed to the combustor, and igniters are energized, this causes ignition to start locally within the combustor, followed by light around all the burners. Finally, in acceleration to idle phase, the fuel mass flow rate is further increased, and the rotational speed increases toward idle value.

## 7.2 Data acquisition and preparation

The required data sets were taken experimentally during several start-up maneuvers and covered the whole operational range of the IPGT during start-up. These data are representative of the operating conditions during start-up and, therefore, they account for all the conditions related to this type of transient maneuver (e.g., bleed valve opening, IGV control, etc.).

The required data for the modeling process were chosen from the entire available data sets for the IPGT that are already categorized and discussed in Chapter 6. The following two maneuvers from cold start-up were considered for making the simulation models:

- The starter is on: 1
- The GT is connected to the grid or not: 1 or 0
- Customer trip does not happen: 0
- The flame is on: 1

The maneuvers can be classified as [1 1 0 1] or [1 0 0 1]. For instance, [1 1 0 1] refers to the situation when the starter is on, the GT is connected to the grid, customer trip does not happen, and the flame is on.

The measured time-series data sets, which are used for training the NARX models are called TR1, TR2, and TR3. They cover the whole operational range of the GT during the start-up procedure. The time step for data acquisition is 1 s.

*Table 7.1* Experimental time-series data sets

| Data sets | Number of data | Operational range of the inputs | | |
|---|---|---|---|---|
| | | $T_{01}$ (K) | $P_{01}$ (kPa) | $M_f$ (kg/s) |
| TR1 | 450 | [289.8; 292.6] | [101.6; 101.9] | [0.73; 4.90] |
| TR2 | 362 | [281.5; 295.9] | [102.3; 102.6] | [0.98; 5.10] |
| TR3 | 510 | [305.4; 308.7] | [101.6; 102.0] | [0.28; 4.80] |
| TE1 | 538 | [295.9; 299.3] | [100.6; 101.0] | [0.37; 4.80] |
| TE2 | 408 | [299.8; 300.9] | [100.6; 101.0] | [0.49; 4.80] |
| TE3 | 397 | [298.7; 299.8] | [100.6; 100.9] | [0.52; 4.80] |

A combination of TR1, TR2, and TR3 was considered for training in such a manner that the resulting model can be confidently generalized for the GT start-up simulation. The data sets TE1, TE2, and TE3 are employed for test and verification of the resulting model. Table 7.1 shows more details about these data and the operational range for the input parameters.

It can be seen that the values of compressor inlet temperature and pressure of the training data sets are different, since they were taken in different seasons (August, October, and December). This choice was made on purpose with the aim to improve the generalization capability of the NARX models. Moreover, the range of variation of $T_{01}$ for TE1, TE2, and TE3 is included in the range of variation of $T_{01}$ of the training data sets. Figure 7.1 shows the trends over time of fuel mass flow rate.

It can be seen that the trends of all maneuvers are similar, but the rate of change of the rotational speed to reach the full-speed/no-load condition is different for each maneuver. In particular, TE1 is very close to TR3 while the trend of the fuel flow for TE2 and TE3 lies in the middle between TR1 and TR3.

## 7.3   GT start-up modeling by using NARX models

NARX has a recurrent dynamic nature and it is commonly used in time-series modeling. NARX includes feedback connections enclosing several layers of the network. Recall that the defining equation of the NARX model can be written as follows [212]:

$$y(t) = f(u(t-1), u(t-2), \ldots, u(t-n_u), y(t-1), y(t-2), \ldots, y(t-n_y))$$

(7.1)

where $y$ is the output variable and u is the externally determined variable. The next value of the dependent output signal $y(t)$ is regressed on previous values of the output signal and previous values of an independent (exogenous) input signal.

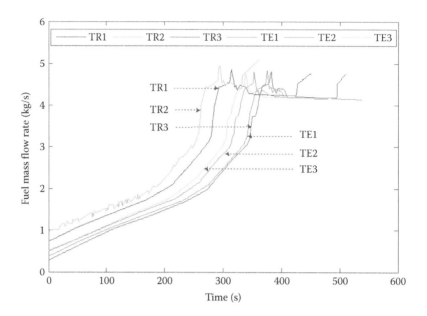

*Figure 7.1* **(See color insert.)** Trend over time of mass flow rate.

As it can be noticed from Equation 6.1, the NARX models developed in this chapter, use the variables at antecedent time steps as inputs. In fact, the exogenous input variable at the current time step $u(t)$ is not an input. This modeling approach allows setting-up and running a software tool in parallel with the GT. This software may also be used for real-time control optimization and GT sensor diagnostics.

The NNT in MATLAB® was employed to build NARX models for a combination of the measured time-series data sets of TR1, TR2, and TR3 in such a manner that the resulting model would cover the whole operational range of the GT start-up operation. The resulting models were obtained after carrying out a thorough sensitivity analysis on NARX parameters (i.e., number of neurons in the hidden layer, number of feedback connections, NARX architecture, and number of delayed time points), in order to obtain the best possible model in terms of accuracy of the trends and *RMSE* for the output parameters. At the same time, the structure of the models was kept as simple as possible by considering the minimum required number of neurons and delayed time points. The models were tested against TE1, TE2, and TE3 maneuver separately.

Figures 7.2 and 7.3 show the closed-loop structure of the NARX models and the block diagram of the complete NARX model used for GT simulation, respectively.

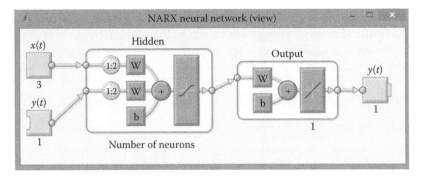

*Figure 7.2* Closed-loop structure of a single NARX model.

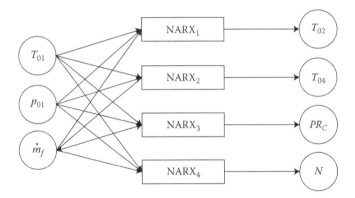

*Figure 7.3* Block diagram of the complete NARX model for IPGT simulation.

As Figure 7.3 shows, the model includes three inputs and four outputs. The inputs are compressor inlet temperature $T_{01}$, compressor inlet stagnation pressure $P_{01}$, and fuel mass flow rate $M_f$. These quantities were selected since they are always available, even in poorly instrumented GTs. The outputs are compressor outlet temperature $T_{02}$, turbine outlet temperature $T_{04}$, compressor pressure ratio $PR_C$, and rotational speed $N$. Figure 7.3 also shows that the complete NARX model has a MIMO structure, such as the model investigated by Bettocchi et al. [145].

The NARX model for each output parameter was trained separately with different number of neurons in order to get the most accurate prediction. Each model was trained by using Levenberg-Marquardt backpropagation (*trainlm*) as the training function; one hidden layer and a tapped delay line with delays from 1 to 2 s at the input. In fact, the NARX model with regressed outputs $y(t-1)$ and $y(t-2)$ at time points $(t-1)$ and $(t-2)$

proved to be most accurate solution, by using a lean structure. With regard to the optimal number of neurons in the hidden layer, the best results for the outputs $T_{02}$, $T_{04}$, $PR_C$, and $N$ were obtained by using 12 neurons in the hidden layer. Values of the measured data and the predictions of NARX models were compared on the basis of *RMSE* already defined according to Equation 4.3 in Chapter 4.

## 7.3.1  NARX model training

Figures 7.4 through 7.7 show the variations of the four output parameters during the GT start-up process of the maneuvers TR1, TR2, and TR3 for the real system (measured data sets) and the trained NARX models.

It should be noted that the training data sets TR1, TR2, and TR3 are supplied as a sequence to the NARX models as required for the training phase. The simulation results in Figures 7.4 through 7.7 were obtained by simulating these maneuvers one by one. It can be observed that the most significant deviations between measured and simulated values occur during the initial phase of the data sets. This means that the NARX models require a time frame of approximately 1 min to stabilize and correctly reproduce the GT behavior (this delay is very clear in Figure 7.12, Figure 7.16, and Figure 7.20 for $T_{04}$).

In general, Figures 7.4 through 7.7 highlight that the NARX models can also follow the physical behavior when the trend remains almost stationary. Moreover, the NARX models tend to smooth the rapid variations, as shown in Figure 7.7 for $T_{04}$.

Figure 7.8 reports the results of the training phase in terms of *RMSE*. To account for the initial delay in the NARX models to work correctly, the values corresponding to the first 10 s of each data set are not used for *RMSE* calculation.

The *RMSE* values slightly depend on the considered training curve with the exception of $N$ for TR3 (*RMSE* equal to 12.0%). The *RMSE* values for $T_{02}$, $T_{04}$, $PR_C$, and $N$ vary in the range 0.7%–4.1%, 0.8%–2.6%, 4.6%–5.6%, and 3.0%–12.0%, respectively. Given that the NARX models are trained with merely three input measurements from experimental data and can generally reproduce the physical behavior, the *RMSE* values were considered acceptable, and the training phase was considered satisfactory.

## 7.3.2  NARX model validation

For validation, the NARX models were tested against three other available experimental time-series data sets, which are indexed as TE1, TE2, and TE3. Figures 7.9 through 7.20 show the results for TE1, TE2, and TE3, respectively. It should be noted that during the simulation phase, the

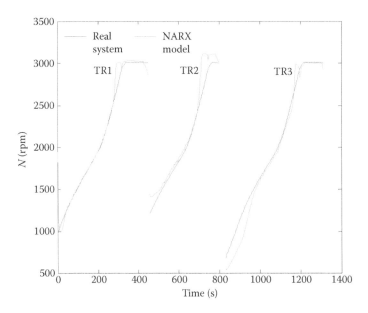

*Figure 7.4* **(See color insert.)** Variations of rotational speed $N$ for the training maneuvers TR1, TR2, and TR3.

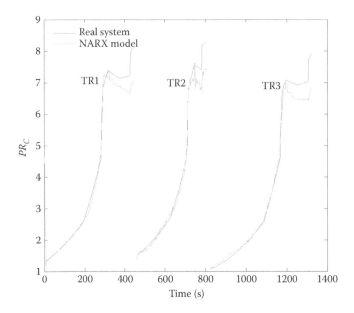

*Figure 7.5* **(See color insert.)** Variations of compressor pressure ratio $PR_C$ for the training maneuvers TR1, TR2, and TR3.

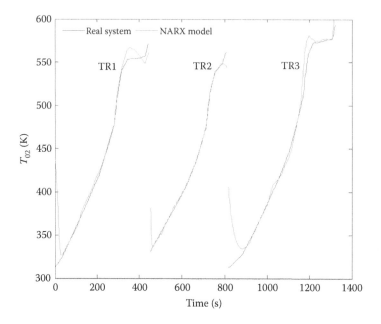

*Figure 7.6* **(See color insert.)** Variations of compress outlet temperature $T_{02}$ for the training maneuvers TR1, TR2, and TR3.

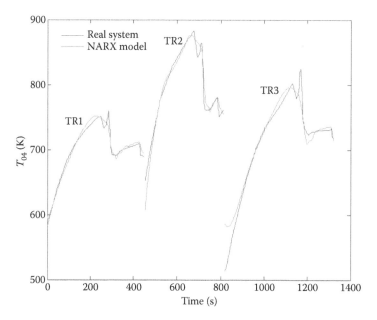

*Figure 7.7* **(See color insert.)** Variations of turbine outlet temperature $T_{04}$ for the training maneuvers TR1, TR2, and TR3.

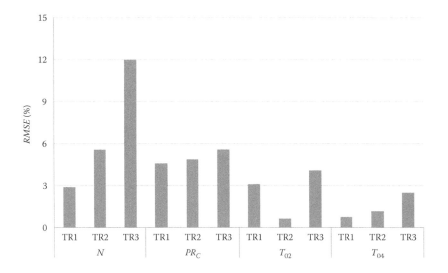

**Figure 7.8** *RMSE* of the NARX models for the training maneuvers TR1, TR2, and TR3.

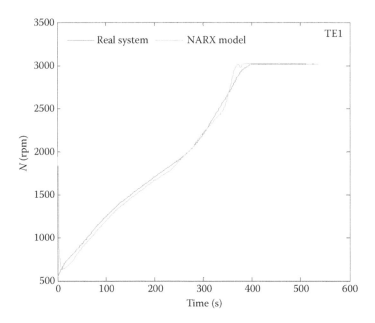

**Figure 7.9** Variations of rotational speed $N$ for the testing maneuver TE1.

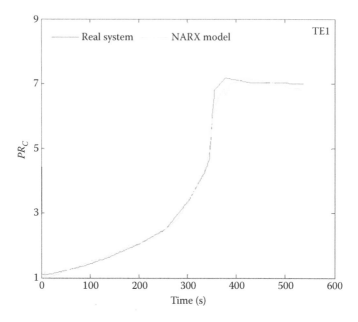

***Figure 7.10*** Variations of compressor pressure ratio $PR_C$ for the testing maneuver TE1.

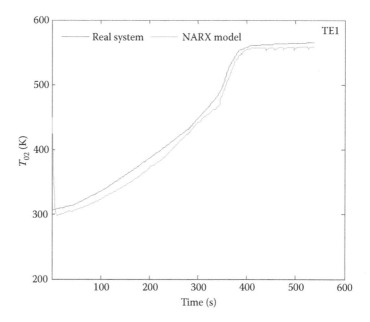

***Figure 7.11*** Variations of compressor outlet temperature $T_{02}$ for the testing maneuver TE1.

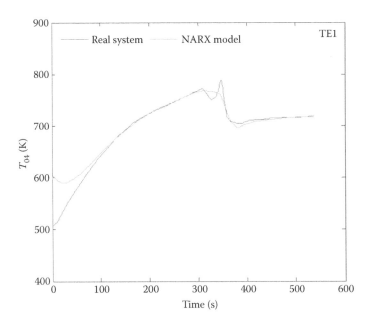

***Figure 7.12*** Variations of turbine outlet temperature $T_{04}$ for the testing maneuver TE1.

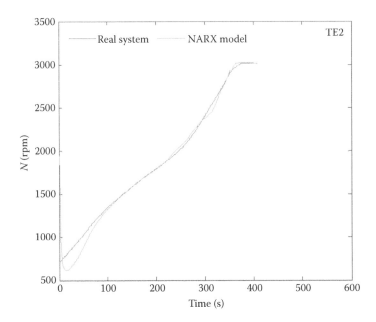

***Figure 7.13*** Variations of rotational speed $N$ for the testing maneuver TE2.

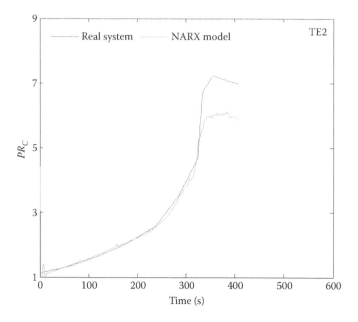

**Figure 7.14** Variations of compressor pressure ratio $PR_C$ for the testing maneuver TE2.

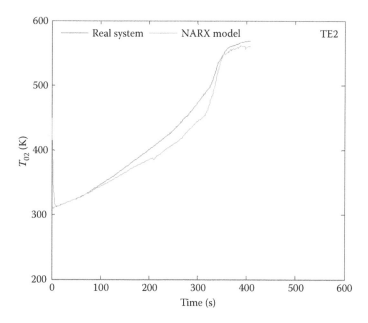

**Figure 7.15** Variations of compressor outlet temperature $T_{02}$ for the testing maneuver TE2.

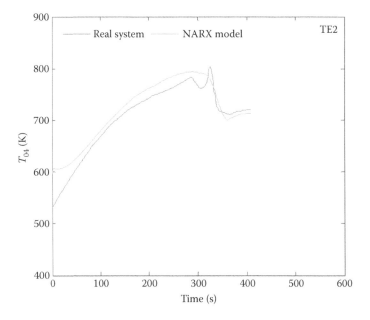

**Figure 7.16** Variations of turbine outlet temperature $T_{04}$ for the testing maneuver TE2.

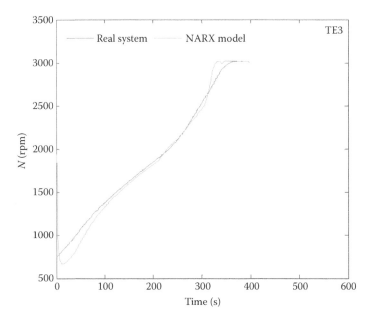

**Figure 7.17** Variations of rotational speed $N$ for the testing maneuver TE3.

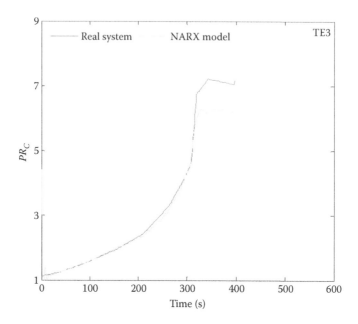

***Figure 7.18*** Variations of compressor pressure ratio $PR_C$ for the testing maneuver TE3.

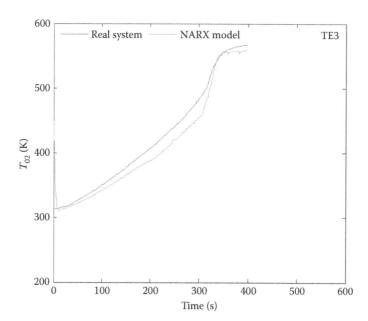

***Figure 7.19*** Variations of compressor outlet temperature $T_{02}$ for the testing maneuver TE3.

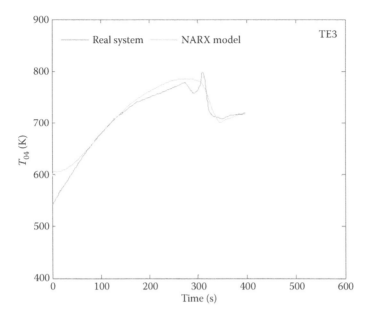

*Figure 7.20* Variations of turbine outlet temperature $T_{04}$ for the testing maneuver TE3.

NARX models are fed with the regressed outputs at time points $y(t-1)$ and $y(t-2)$ estimated by the NARX model itself at antecedent time steps.

It can be seen that, in all cases, the trends of the real system and the NARX models are very similar. This means that the NARX models can follow the changes in GT parameters, even though they are subject to significant changes. In fact, as an example, when the rotational speed is varied from 500 rpm to 3000 rpm in approximately 6 min consequently, the compressor pressure ratio increases from about 1 to about 7.

At the same time, the NARX models can also reproduce fewer significant changes, as, for instance, can be observed in the trends of turbine outlet temperature $T_{04}$. Finally, the stable operation can also be reproduced very satisfactorily, as it can be clearly seen during the last minutes of each transient maneuver (in particular, in Figures 7.9 through 7.12). As it can be seen, there are just two cases with a noticeable deviation of the measured and predicted trends, that is, the compressor pressure ratio $PR_C$ for TE2 and TE3 (see Figures 7.14 and 7.18). In any case, as shown below, the overall deviation can still be acceptable. Therefore, it can be concluded that the NARX models reproduced the three testing transients TE1, TE2, and TE3 with a good accuracy. Figure 7.21 summarizes the results in terms of *RMSE* for the testing maneuvers. Also, in this case, to account for the initial delay of the NARX model to work correctly, the

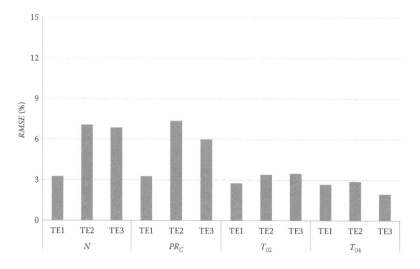

**Figure 7.21** *RMSE* of the NARX models for the testing maneuvers TE1, TE2, and TE3.

values corresponding to the first 10 s of each data set are not used for *RMSE* calculation.

A different behavior can be observed for compressor and turbine outlet temperatures, compared to pressure ratio and rotational speed. In fact, *RMSE* values for temperatures are always lower than 3.5%, approximately. Although the overall trend is reproduced almost correctly, the *RMSE* maximum values of $PR_C$ and $N$ are almost twice (7.4% for $PR_C$ and 7.1% for $N$). As observed for the training data sets, the *RMSE* values slightly depend on the considered training curve.

In conclusion, the results show that the NARX models have the potential to simulate and predict GT dynamic behavior. However, Figures 7.12 and 7.21 highlight that general guidelines about the order of magnitude of the errors are difficult to draw, since they may change as a function of the considered maneuver and measurable quantity. The results reported here represent a good compromise on the NARX model prediction capability of the four selected output variables. It has to be considered that the comparison to experimental data was mainly intended to evaluate the agreement of the trends rather than the numerical values. According to the modeling hypotheses, the structure of the NARX models was kept as simple as possible so that only three usually available variables were supplied as inputs. Moreover, the differences in the numerical values can also be attributed to the accuracy of the GT experimental measurement system.

## 7.4   Summary

In this chapter, the dynamic behavior of a heavy-duty single-shaft GT during the start-up phase is investigated. For this reason, NARX models of the IPGT were constructed by using three measured time-series data sets. The resulting NARX models were tested against three other available experimental data sets for verification of the models.

For this purpose, four important outputs from the models and their corresponding values from the measured data sets were compared (compressor and turbine outlet temperature, compressor pressure ratio and rotational speed as a function of compressor inlet temperature, and pressure and fuel mass flow rate). According to the results, the NARX models have the capability of capturing and predicting GT dynamics during start-up. In most cases, the deviation between measured and simulated values is acceptable (e.g., lower than approximately 3.5% for compressor and turbine outlet temperatures), but it can also increase to nonnegligible values for compressor pressure ratio and rotational speed (maximum deviations equal to 7.4% and 7.1%, respectively). In general, the physical behavior is well captured by the NARX models and the influence of the considered data set is negligible.

One of the strong points of this methodology is the simplicity of the developed NARX models. It is clear that accurate modeling (e.g., through a physics-based approach) does need much information about the bleed flows and IGV control which are usually unknown as they are confidential manufacturer's data or simply unavailable. For this reason, and in order to overcome this lack of information, NARX models were employed as a black-box tool to model the GT for the whole range of start-up operation. The resulting NARX models can reproduce a very complicated and usually difficult-to-model unsteady behavior and can capture system dynamics with acceptable accuracy. It was shown that NNs could be considered a reliable alternative to conventional methods of system identification and modeling.

The results of this modeling approach, which uses only the variables at antecedent time steps as inputs (i.e., no information about the current time step is required), allow the set-up of a powerful and easy-to-build simulation tool which may be used for real-time control and sensor diagnostics of GTs.

*chapter eight*

# Design of neural network-based controllers for GTs

True creativity often starts where language ends.

**Arthur Koestler**
*Hungarian-British author and journalist, 1905–1983*

Modeling of control systems before their implementation in real plants is an efficient and cost-saving strategy in industrial applications. The need for controllers with high-quality standards to reliably manipulate operations in complex industrial systems has been increasing remarkably. These controllers should have the capability of dealing with restrictions on control strategies and internal variables [170]. This necessity has led to the development of different kinds of controllers, which can be successfully applied to industrial plants. However, because of the nonlinear nature of industrial systems and deviation of control systems from the design objectives, there are still high demands for controllers and control approaches which can incorporate system nonlinearity. ANNs have a high capability in modeling and control of dynamic systems such as GTs.

In this chapter, the structures of a conventional PID controller and ANN-based controllers including MPC and feedback linearization control (NARMA-L2) are briefly described. Their related parameters will be set up according to the requirements of the controller design for a single-shaft GT. Finally, a comparison is made of the performances of these controllers.

## 8.1 GT control system

The GT system which is used for controller design purposes in this chapter is a nonlinear dynamic model of a low-power single-shaft GT which was developed and verified for loop-shaping control purposes [27]; and was already discussed and simulated in Chapter 5. It is employed for designing of PID and ANN-based controllers in this chapter. Figure 8.1 shows the closed-loop diagram of the control system for the GT engine system. It includes the plant that is the GT system, the controller, random reference, and indicator blocks. Fuel mass flow rate and rotational speed

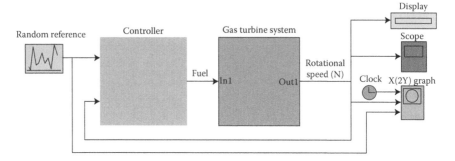

*Figure 8.1* (**See color insert.**) The closed-loop diagram of the control system for the GT engine system.

are input and output of the plant, respectively. The controller could be any of the controller structures including MPC, NARMA-L2, or PID, as will be discussed later in this chapter. They are already implemented in MATLAB® software, and their parameters need to be tuned according to the control requirements.

Figure 8.2 shows the random reference block diagram with the adjusted values. The aim is to maintain the rotational speed at a constant value of 700 rpm (set point) when the input of the control system changes with the random reference (step function). The random reference is adjusted between 695 and 705 rpm with a sample time of 0.2 s. It produces random step functions based on the adjusted parameters.

The objective of the controller design is to achieve a satisfactory response (to the reference input) for each of the controllers with the following conditions:

- Rise time < 0.5 s
- Settling time < 2 s
- Maximum overshoot < 15%
- Steady-state error < 5%

Figure 8.3 shows a typical response to a standard test signal (input) which is usually a step function (random reference) with its characteristics including rise time $(T_r)$, settling time $(T_s)$, peak time $(T_p)$, maximum overshoot $(M_p)$, and steady-state error $(e_{ss})$. The following provides a short definition of these terms:

- Rise time: The time required for the response signal to rise from 10% to 90% of the final value (its set-point value).
- Settling time: The time elapsed for the response signal to get and remain within an error band (±5%) of the final value.

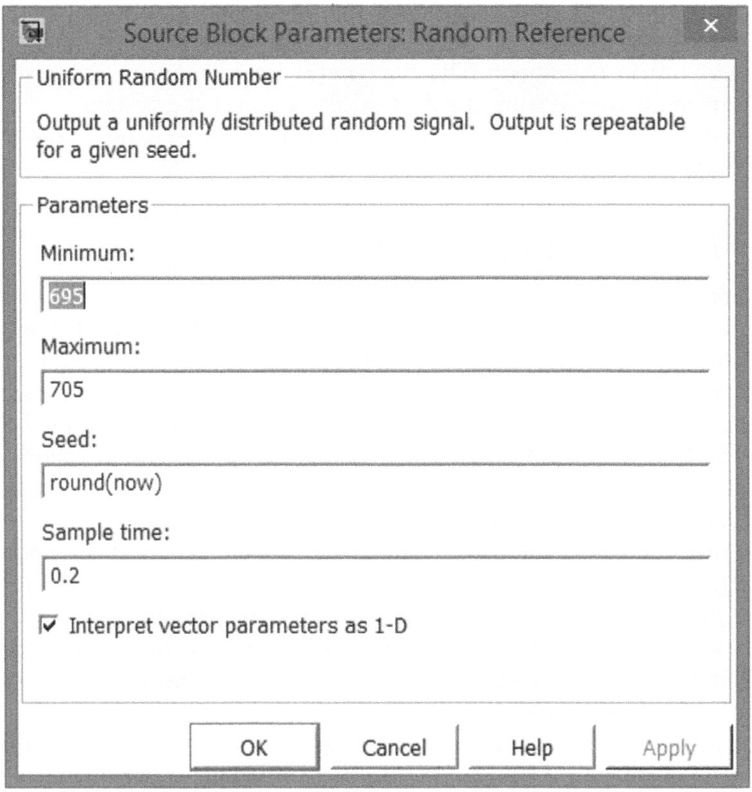

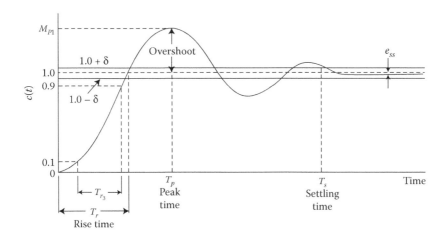

*Figure 8.2*  Random reference block diagram.

*Figure 8.3*  A typical response to a standard test signal with its characteristics.

- Peak time: The time elapsed for the response signal to reach its first maximum value.
- Maximum overshoot: The maximum peak value of the response curve measured from the desired response of the system.
- Steady-state error: The error that remains after transient conditions disappear in a control system.

## 8.2   Model predictive controller

MPC pioneered by Richalet et al. [104] and Cutler et al. [234] has been widely used in a variety of process plants all around the world. The most important benefits of MPC, which have made it successful in industrial applications, include the capability of handling structural changes, non-minimal phase, unstable processes, as well as multivariable control problems [235]. Besides, MPC is sufficiently fast for online computations and can take account of actuator limitations. It is an easy-to-tune method and can operate closer to constraints.

Figure 8.4 shows the basic structure of MPC [235]. As it can be seen from this figure, the model predicts output of the system based on the future inputs, past inputs, and past outputs. Output is compared to a reference value and the difference (error) goes to the optimizer, which determines the future inputs for the model. Optimization process occurs based on the system constraint and a predefined cost function.

In an ANN-based MPC, an NN represents the forward nonlinear dynamics of the plant. It is used to predict future plant performance. The function of the controller is to calculate the control input that optimizes plant performance over a specified future time horizon.

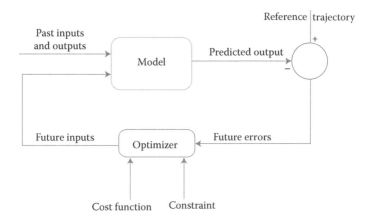

***Figure 8.4*** A basic MPC structure. (From P. Orukpe, *Basics of Model Predictive Control,* Imperial College, London, 2005, 27 pp.)

The receding horizon technique presented by Soloway et al. [236] is the basis of MPC methodology. According to this methodology, the predictions of the plant response over a specified time horizon made by the NN model are employed by a numerical optimization program to determine the control signal that minimizes the following performance criterion over the specified horizon [212]; Equation 8.1 shows the mathematical description of the MPC process.

$$J = \sum_{j=N_1}^{N_2}(y_r(t+j) - y_m(t+j))^2 + \sum_{j=1}^{N_u}(u'(t+j-1) - u'(t+j-2))^2 \quad (8.1)$$

where $N_1$, $N_2$, and $N_u$ represent the horizons over which the tracking error and the control increments are evaluated. $u'$, $y_r$, and $y_m$ are the tentative control signal, the desired response, and the network model response, respectively. The $\rho$ value determines the contribution that the sum of the squares of the control increments has on the performance index. This process is also illustrated in Figure 8.5 [212]. As this figure shows, the controller, which has already been implemented in Simulink®, consists of the NN plant model and the optimization block [212]. The optimization block determines the values of $u'$ that minimize $J$, and then the optimal $u$ is the input to the plant [212].

## 8.2.1 Design of ANN-based MPC

The first step in MPC design process is to determine the NN plant model (system identification). Then, the plant model is used by the controller to

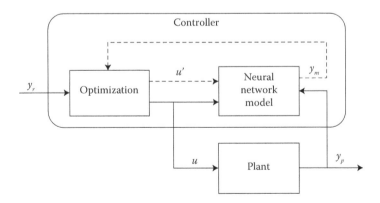

*Figure 8.5* NN-based model predictive controller. (From M. H. Beale, M. T. Hagan, and H. B. Demuth, Neural Network Toolbox™ User's Guide, R2011b ed., Natick, MA: MathWorks, 2011, 404 pp.)

predict future performance [212]. NN is trained using the NN training signal, which is the prediction error between the plant output and the NN output. Previous plant outputs and previous inputs are employed by the NN plant model to predict future values of the plant output. Figure 8.6 shows the training process flowchart of the NN plant model. This network has been implemented in NN Toolbox software of MATLAB and can be trained offline using different training algorithms for the operational data sets obtained from the plant [212].

The closed-loop diagram of the control system for the GT engine system with the MPC is similar to Figure 8.1 when the controller block is replaced by the NN predictive controller block shown in Figure 8.7. This block is already implemented in Simulink/MATLAB. Design of

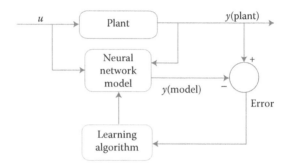

*Figure 8.6* Training process in an ANN-based MPC. (From M. H. Beale, M. T. Hagan, and H. B. Demuth, Neural Network Toolbox™ User's Guide, R2011b ed., Natick, MA: MathWorks, 2011, 404 pp.)

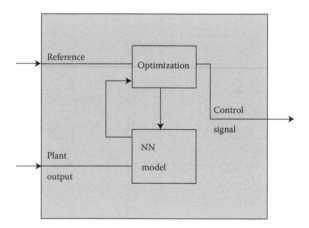

*Figure 8.7* ANN-based model predictive control block. (From M. H. Beale, M. T. Hagan, and H. B. Demuth, Neural Network Toolbox™ User's Guide, R2011b ed., Natick, MA: MathWorks, 2011, 404 pp.)

ANN-based MPC in MATLAB environment includes different steps, which will be explained in the following sections.

### 8.2.1.1   System identification of ANN-based MPC

Before the controller is designed, system identification process should be completed, and the NN plant model must be developed. The optimization algorithm employs these predictions to determine the control inputs that optimize future performance. Figure 8.8 shows the block diagram of plant identification for CSGT system with all the adjusted parameters for generating data and training the NN model of the system.

As it can be seen in Figure 8.8, minimum and maximum values for the plant input (mass fuel rate) are 0.00367 and 0.027 kg/s. For the plant output (rotational speed), minimum and maximum values are 650 and 733 rpm, respectively. Before the NN training stage, 8000 data sets for the GT input and output were generated by considering the minimum and maximum interval values as 0.2 s and 0.8 s, respectively. These data were generated using the "Generate Training Data" option. The integrated program can

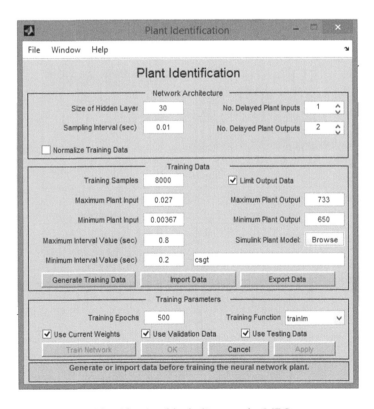

*Figure 8.8* GT system identification block diagram for MPC.

generate training data by applying a series of random step inputs to the Simulink plant model. The size of the hidden layer and the number of delayed plant inputs and outputs were adjusted at 30 s and 2 s, respectively. The sampling interval was fixed at 0.02 s. The training proceeds according to the selected training function (*trainlm*).

After the training is complete, the response of the resulting plant model is displayed, as it is shown in Figure 8.9. Separate plots for validation data are shown in Figure 8.10. As it can be seen from Figures 8.9 and 8.10, the results of training for NN model of the GT are satisfactory.

### 8.2.1.2 Adjustment of controller parameters for ANN-based MPC

After the system identification process is completed, the model predictive controller is designed. Figure 8.11 shows NN predictive control block diagram with its adjusted parameters. In this figure, the controller horizons $N_2$ and $N_u$ have been tuned at 7 and 2, respectively. $N_1$ is fixed at the value 1 by default. The weighting factor $\rho$ and the search parameter $\alpha$ have been adjusted at 0.05 and 0.01, respectively. The task of parameter $\alpha$ is to control the optimization by specifying how much reduction in performance

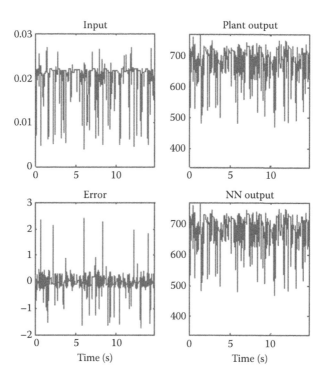

*Figure 8.9* Training data for NN predictive control.

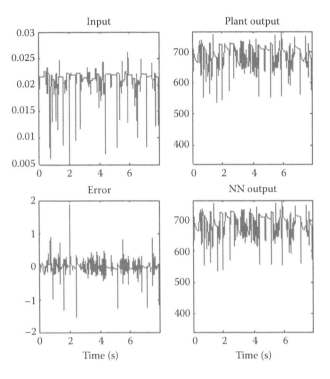

*Figure 8.10* Validation data for NN predictive control.

is required for a successful optimization step. The number of iterations of the optimization algorithm at each sample time has been tuned at 2 s. Besides, different linear minimization routines can be used by the optimization algorithm. *csrchbac* is the best selected minimization routine for this design.

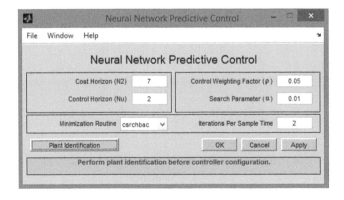

*Figure 8.11* NN predictive control block diagram.

## 8.2.2    Simulation of ANN-based MPC

Simulation is the last stage of ANN-based control design after adjustment of the controller parameters. Now, the closed-loop control system, shown in Figure 8.1, can be run to simulate the whole system. The result of the simulation is shown in Figure 8.12.

# 8.3    Feedback linearization controller (NARMA-L2)

The nonlinear autoregressive moving average (NARMA) model is a standard model that is employed to represent general discrete-time nonlinear systems. The NARMA model represents input–output behavior of finite-dimensional nonlinear discrete-time dynamical systems in a neighborhood of the equilibrium state [237]. However, it is not suggested for adaptive control purposes using NNs because of its nonlinear dependence on the control input [237]. Equation 8.2 indicates the mathematical description of the NARMA.

$$y(k + d) = f[y(k), y(k - 1),\ldots y(k - n + 1),$$
$$u(k), u(k - 1), \ldots u(k - n + 1)]$$
(8.2)

where $u(k)$ is the system input and $y(k)$ is the system output. An NN is needed to be trained to approximate the nonlinear function $f$ for the system identification stage. Because the NARMA model described by Equation 8.2 is slow, an approximate model is used to represent the system. This

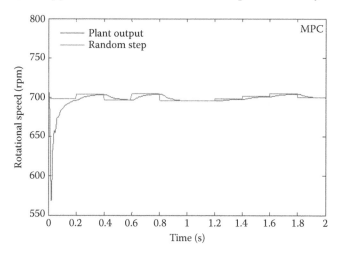

*Figure 8.12* Response of GT system with MPC to random step inputs.

model called NARMA-L2 can be described mathematically according to Equation 8.3, where $d \geq 2$.

$$
\begin{aligned}
y(k + d) = \, &f[y(k), \, y(k - 1), \ldots y(k - n + 1), \, u(k), \, u(k - 1), \\
&\ldots u(k - n + 1)] + g[y(k), \ldots y(k - n + 1), \, u(k), \\
&\ldots u(k - n + 1)] \cdot u(k + 1)
\end{aligned}
\tag{8.3}
$$

The corresponding controller for NARMA-L2 model is mathematically defined according to Equation 8.4, which is realizable for $d \geq 2$.

$$
u(k + 1) = \frac{y_r(k + d) - f[y(k), \ldots y(k - n + 1), \, u(k), \ldots u(k - n + 1)]}{g[y(k), \ldots y(k - n + 1), \, u(k), \ldots u(k - n + 1)]}
\tag{8.4}
$$

Figure 8.13 shows a block diagram of NARMA-L2 controller with approximation functions $f$ and $g$, and the time delays TDL, all implemented in the NARMA-L2 control block. Controller is a multilayer NN that has been successfully applied in the identification and control of dynamic systems [212]. The main idea behind the NARMA-L2 is transforming nonlinear system dynamics into linear dynamics. It is a rearrangement of the NN plant model, which is trained offline.

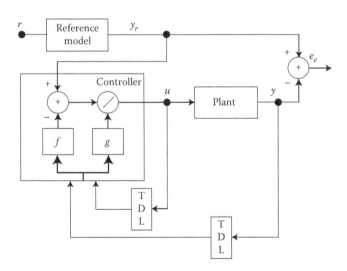

*Figure 8.13* A block diagram of NARMA-L2 controller. (From M. H. Beale, M. T. Hagan, and H. B. Demuth, Neural Network Toolbox™ User's Guide, R2011b ed., Natick, MA: MathWorks, 2011, 404 pp.)

## 8.3.1   Design of NARMA-L2

NARMA-L2 controller block has already been implemented in Simulink–MATLAB. There are two main steps in using NARMA-L2 including system identification and control design. At the system identification stage, an NN model of the plant is developed. This stage that includes the block diagram representation of the system identification and the training process is similar to system identification of MPC, described earlier in this chapter. The closed-loop diagram of the control system for the GT engine system with the NARMA-L2 controller is also similar to Figure 8.1, when the controller block in this figure is replaced by NARMA-L2 controller block which is shown in Figure 8.14.

Figure 8.15 shows the block diagram of plant identification for the CSGT, which uses the NARMA-L2 controller with all the adjusted parameters for generating data and training the NN model of the system.

As it can be seen from this figure, minimum and maximum values for the plant input (fuel mass flow rate) are 0.00367 and 0.027 kg/s. For the plant output (rotational speed), minimum and maximum values are 650 and 733 rpm, respectively. Before the NN training stage was performed; 10,000 data sets for the GT input and output were generated by considering the minimum and maximum interval values as 0.1 s and 1 s. These data were generated using the option Generate Training Data. The integrated program can generate training data by applying a series of random step inputs to the Simulink model of the plant. The size of the hidden layer

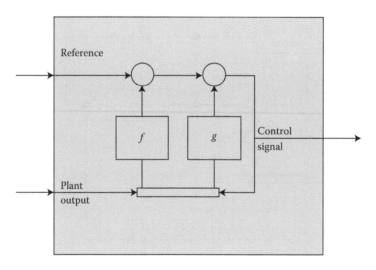

*Figure 8.14* NARMA-L2 control block. (From M. H. Beale, M. T. Hagan, and H. B. Demuth, Neural Network Toolbox™ User's Guide, R2011b ed., Natick, MA: MathWorks, 2011, 404 pp.)

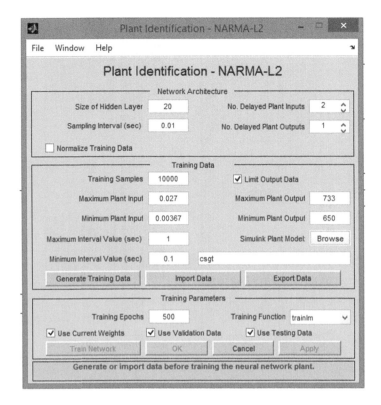

*Figure 8.15* GT system identification block diagram for NARMA-L2.

and the number of delayed plant inputs and outputs were adjusted at 20, 2, and 1, respectively. The sampling interval was fixed at 0.01 s. The training proceeded according to the selected training function (*trainlm*). After the completion of the training, the response of the resulting plant model was displayed, as shown in Figure 8.16. Separate plots for validation data are shown in Figure 8.17. As it can be seen from Figures 8.16 and 8.17, the results of training for NN model of the GT are satisfactory.

## 8.3.2 Simulation of NARMA-L2

Simulation is the last stage of NARMA controller design. At this stage, the closed-loop control system can be run to simulate the whole system. The result of the simulation is shown in Figure 8.18. The result shows that NARMA-L2 controller can accurately follow the value and trend of changes in the system input. Besides, the reaction of the controller to the changes is very fast.

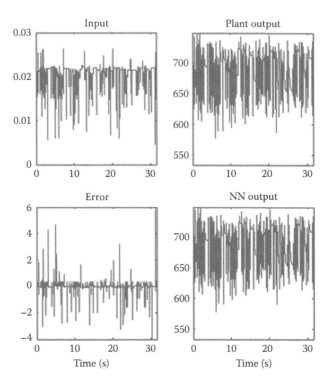

***Figure 8.16*** Training data for NARMA-L2 controller.

## *8.4   PID controller*

PID controller was first introduced in the industry in 1939 and has remained to be the most widely used controller in industrial control systems until today [238]. PID is a generic feedback control system that acts based on the difference between values of a measured process variable and a desired setpoint. The controller objective is to minimize this difference, which is called error, by adjusting the process control inputs. The PID controller includes the proportional ($P$), the integral ($I$), and the derivative ($D$) values that can be interpreted in terms of time. $P$, $I$, and $D$ depend on the present error, accumulation of past errors, and prediction of future errors, respectively, [238]. By tuning these three parameters, the controller can provide the required control action designed for a specific process. Based on the application, it is also common to use just PI, PD, $P$, or $I$ controllers. The popularity of PID controllers is specifically because of their flexibility for giving the designer a larger number of design options

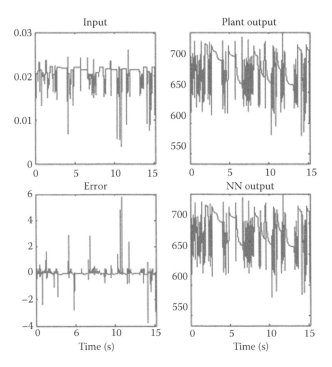

*Figure 8.17* Validation data for NARMA-L2 controller.

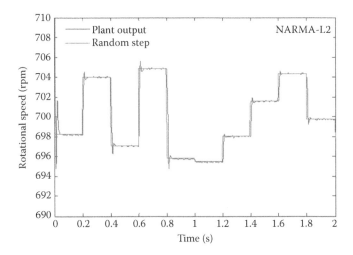

*Figure 8.18* **(See color insert.)** Response of GT system with NARMA-L2 control-ler to random step inputs.

on the basis of the system dynamics. The PID algorithm is described by Equation 8.5

$$u(t) = K\left( e(t) + \frac{1}{T_i}\int_0^t e(\tau)d(\tau) + T_d\frac{de(t)}{d(t)} \right) \tag{8.5}$$

where $y$ is the measured process variable, $r$ is the reference variable, $u$ is the control signal, and $e$ is the control error. The controller parameters are proportional gain $K$, integral time $T_i$, and derivative time $T_d$. The control signal, thus is a sum of three terms including $P$, $I$, and $D$. The reference variable is often called the set point [239]. Figure 8.19 shows the block diagram of a PID controller operating in an in-series path with the plant, as it is used in this chapter [240].

### 8.4.1   Design of PID controller

PID controller block has been implemented in the Simulink–MATLAB and its gains are tunable either manually or automatically according to the PID tuning algorithm in MATLAB. The closed-loop diagram of the control system for the GT engine system with the PID controller is also similar to Figure 8.1, when the controller block is replaced by PID controller block which is shown in Figure 8.20.

The objective of tuning the PID gains is to achieve a good balance between performance and robustness while keeping the closed-loop stability. Therefore, the tuning is performed in a way that the closed-loop system tracks reference changes, suppresses disturbances as rapidly as possible, and its output remains bounded for bounded input. Besides, the loop design should have enough gain margin and phase margin to allow

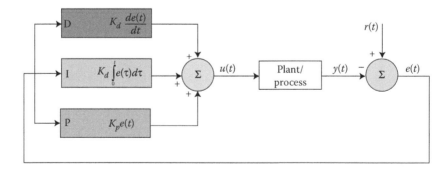

***Figure 8.19*** Block diagram of a PID controller in an in-series path with the plant. (From Wikimedia Commons 2013. [Online]. Available: http://commons.wikimedia. org. [Accessed 2013].)

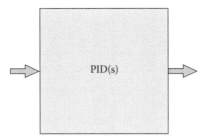

**Figure 8.20** PID control block in MATLAB®. (From M. H. Beale, M. T. Hagan, and H. B. Demuth, Neural Network Toolbox™ User's Guide, R2011b ed., Natick, MA: MathWorks, 2011, 404 pp.)

for modeling errors or variations in system dynamics. According to the algorithm, at the first stage of the tuning, an initial controller is designed by choosing a bandwidth to achieve the balance between performance and robustness based upon the open-loop frequency response of the linearized model. When the response time, bandwidth, or phase margin is interactively changed using the PID tuner interface, the new PID gains are computed by the algorithm. This process continues until the desirable PID controller is achieved [212]. According to the algorithm, Equation 8.5 can be rewritten as follows:

$$u = P + I\frac{1}{s} + D\frac{N}{1 + N\,1/s} \tag{8.6}$$

where $P$, $I$, and $D$ are proportional, integral, and derivative gains, respectively. $N$ is a filter coefficient. Table 8.1 shows the values of the tuned PID gains. Figure 8.21 shows the PID control algorithm block with the tuned PID gains for the GT engine system.

## 8.4.2 Simulation of PID controller

After the completion of PID controller design, the closed-loop control system can be run to simulate the whole system. The result of the simulation is shown in Figure 8.22. As it can be seen from this figure, the response

*Table 8.1* Tuned PID gains for the GT engine

| PID element | Tuned gain value |
|---|---|
| $P$ | 1.085496262366938e-04 |
| $I$ | 0.008414700379816 |
| $D$ | 1.023227413576635e-07 |
| $N$ | 2.173913000000000e+02 |

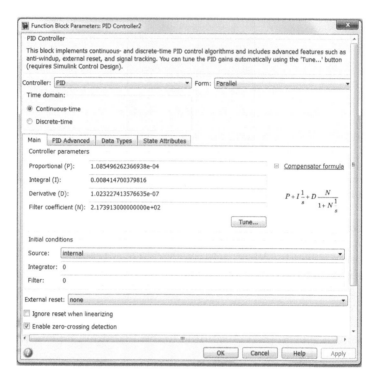

*Figure 8.21* PID control algorithm block for tuning PID gains.

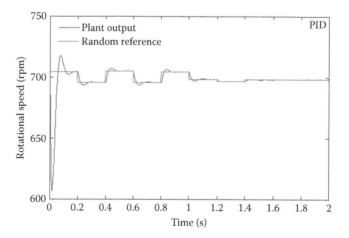

*Figure 8.22* **(See color insert.)** Response of GT system with PID controller to random step inputs.

of the controller to the changes of the system input is fast, and after about 0.2 s, it stabilizes and follows the value and trend of the changes.

## 8.5 Comparison of controllers performance

To compare the results of performances of all three designed controllers, they were run in a common Simulink environment with the same input for their control systems. Figure 8.23 shows the resulting Simulink model including the designed ANN-based MPC, feedback linearization controller (NARMA-L2), and conventional PID controller. Specifications of the random reference (step function) and the GT system for all controllers are the same, as already discussed in this chapter. The simulation was run for 2 s, which was long enough time for capturing the complete dynamics of all the three controllers. Figure 8.24 shows the performances of the controllers. Figures 8.25 and 8.26 show the same performances from closer perspectives to the set point of rotational speed (700 rpm) and the initial response, respectively.

As can be seen from Figures 8.24 through 8.26, all three controllers satisfied the controller design objectives. However, NARMA-L2 controller has a superior performance compared to MPC and PID. It follows the value and trend of the changes faster and more accurately. The settling time, rise time, maximum overshoot, and maximum steady-state error for the response of NARMA-L2 is considerably less than the corresponding values for the other controllers.

As shown in Figure 8.26, the step response of the GT system with each of the controllers starts with an *undershoot*. This is because the GT is a nonminimum phase (NMP) system. This is explained in Section 8.6 of this chapter.

## 8.6 NMP systems

From a controller point of view, all systems can be divided into three main groups based on their phase response and locations of the poles and zeros of their transfer functions in the complex plane also called $S$ plane. This classification consists of minimum phase (MP), pass, and NMP systems. When all poles and zeros of the transfer function of a system are located in the left half of the complex plane, it is called an MP system. In this case, the poles and zeros have negative real parts. In a pass system, the transfer function has a pole-zero pattern which is antisymmetric about the imaginary axis. An NMP system is the one whose transfer function has one or more poles or zeros in the right half of the complex plane. Figures 8.27 and 8.28 show the step response of a typical type of NMP and MP systems, respectively [241].

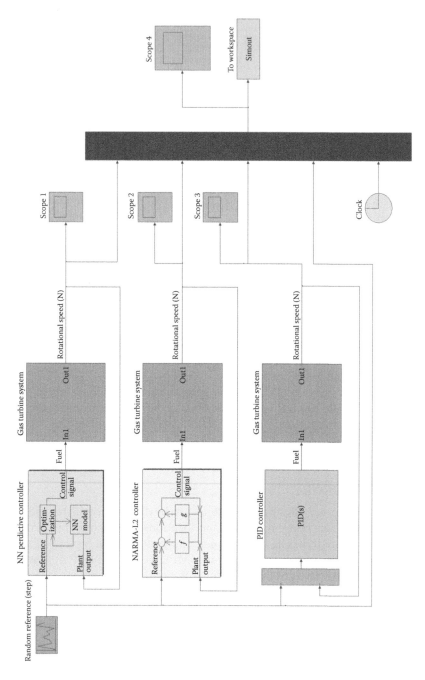

*Figure 8.23* **(See color insert.)** Simulink® model of the ANN-based MPC, NARMA-L2, and PID controllers for a single-shaft GT.

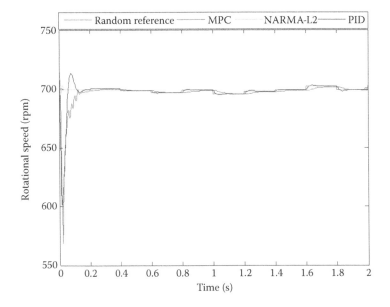

***Figure 8.24*** **(See color insert.)** Performances of three different controllers for a single-shaft GT.

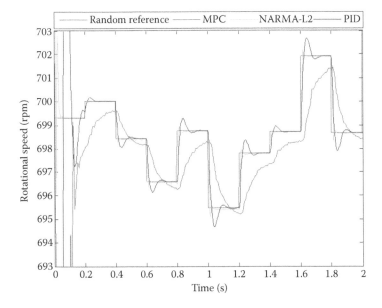

***Figure 8.25*** **(See color insert.)** A close-up perspective of the performances of three different GT controllers.

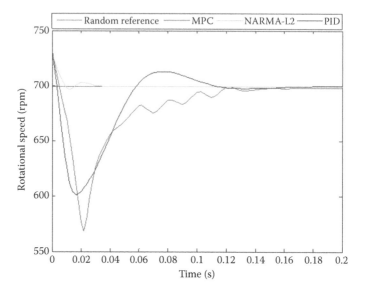

***Figure 8.26*** **(See color insert.)** A close-up perspective of the initial responses of three different GT controllers.

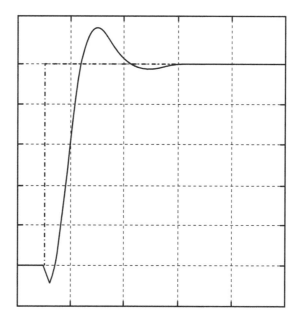

***Figure 8.27*** Step response of an NMP system. (From D. Viswanath, Study of Minimum and Non-Minimum Phase Systems, 2009. [Online]. Available: http://www.scribd.com/doc/16067103/study-of-nonminimum-phase-systems. [Accessed January 10, 2014].)

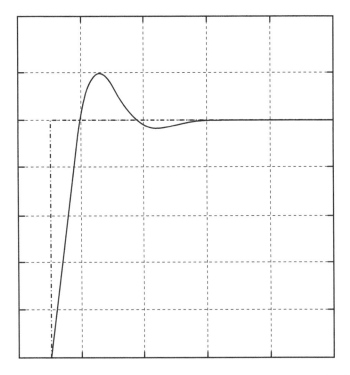

*Figure 8.28* Step response of an MP system. (From D. Viswanath, Study of Minimum and Non-Minimum Phase Systems, 2009. [Online]. Available: http://www.scribd.com/doc/16067103/study-of-nonminimum-phase-systems. [Accessed January 10, 2014].)

As it can be seen from Figure 8.27, an NMP system behaves faultily at the start of the response with an *undershoot*. The output becomes first negative before changing direction and converging to its positive steady-state value. This kind of behavior, which makes the response slow, could arise due to the time delay in the system. Some of the system identification techniques do not take into account time delay and approximate the system as NMP. NMP systems may face internal stability issues, which can be fixed using appropriate controllers.

NMP phenomenon has been already observed in GT systems [242]. The *undershoot* observed in Figure 8.26 is because the GT system is an NMP system. However, as it can be seen from the figures, the controllers could quickly and successfully correct this behavior and bring the system to a stable situation. The reaction of NARMA-L2 as an ANN-based controller to the faulty behavior is quicker than the PID controller. NARMA-L2 shows again a superior performance compared with other controllers in this case. Its settling time, rise time, and maximum overshoot are less than those in conventional PID controller. The previous experience and research results demonstrate that if the response is well controlled by designing a suitable controller, then NMP phenomenon does not cause any problem for the operation of GTs [242].

## 8.7  Summary

This chapter presented three different controller structures for a low-power single-shaft GT. These controllers are ANN-based model predictive controller, ANN-based feedback linearization controller (NARMA-L2), and conventional PID controller. They were briefly described, and their parameters were adjusted and tuned in Simulink–MATLAB environment according to the requirement of the GT system and the control objective. Finally, performances of the controllers were explored and compared. The results show that NARMA-L2 has a superior performance than the other investigated controllers. The settling time, rise time, and maximum overshoot for the response of NARMA-L2 are less than the corresponding ones for the conventional PID controller.

# References

1. R. T. Harman, *Gas Turbine Engineering: Applications, Cycles and Characteristics*, London, UK: Macmillan, 1981, 270 pp.
2. M. P. Boyce, *Gas Turbine Engineering Handbook*, 4th ed., Oxford: Butterworth-Heinemann Ltd, Elsevier Science & Technology, 2011, 1000 pp.
3. G. G. Kulikov and H. A. Thompson, *Dynamic Modelling of Gas Turbines*, London Ltd: Springer, 2010, 336 pp.
4. J. W. Sawyer, *Sawer's Gas Turbine Engineering Handbook*, Stamford, Connecticut: Gas Turbine Publications, 1966, 552 pp.
5. A. M. Y. Razak, *Industrial Gas Turbines: Performance and Operability*, 1st ed., Cambridge, England: Woodhead Publishing Limited, 2007, 602 pp.
6. H. Asgari, X. Q. Chen, and R. Sainudiin, Modelling and simulation approaches for gas turbine system optimization, in Maki K. Habib, and J. Paulo Davim (eds.), *Engineering Creative Design in Robotics and Mechatronics*, Hershey, PA: IGI Global, 2013, pp. 240–264.
7. G. V. Beard and M. R. Rollins, Engineering economics, in *Power Plant Engineering*, London, UK: Springer Science & Business Media Inc., 1996.
8. Wikimedia Commons Webpage, 2013. [Online]. Available: http://en.wikipedia.org/wiki/File:Brayton_cycle.svg. [Accessed 14 June 2013].
9. H. Cohen, G. Rogers, and H. Saravanamut, *Gas Turbine Theory*, 5th ed., Harlow, England: Pearson Education, 2001.
10. H. Asgari, X. Q. Chen, and R. Sainudiin, Considerations in modelling and control of gas turbines—A review, in *The 2nd International Conference on Control, Instrumentation, and Automation (ICCIA)*, Shiraz, Iran, December 2011, pp. 84–89.
11. Wikimedia Commons, 2012. [Online]. Available: http://commons.wikimedia.org.
12. THM Gas Turbine Basic Training, *Turbo Training Catalogues*, Germany: MAN Diesel & Turbo Co., 2009.
13. W. P. J. Visser, O. Kogenhop, and M. Oostveen, A generic approach for gas turbine adaptive modelling, *ASME Journal of Engineering for Gas Turbines and Power*, vol. 128, no. 1, pp. 13–19, 2006.
14. L. Ljung and T. Glad, *Modelling of Dynamic Systems*, Englewood Cliffs, New Jersey: PTR Prentice Hall, Prentice Hall Information and System Sciences Series, 1994, 361 pp.
15. T. Giampaolo, *Gas Turbine Handbook – Principles and Practice*, 4th ed., Lilburn, Georgia, USA, Fairmont Press Inc., 2009, 447 pp.

16. D. Clifton, *Condition Monitoring of Gas Turbine Engines*, University of Oxford, London, UK, 2006.
17. Y. K. Lee, M. Yuan, T. Fishe, D. N. Mavris, and V. Volovoi, A fault diagnosis method for industrial gas turbines using Bayesian data analysis, *ASME Journal of Engineering for Gas Turbines and Power*, vol. 132, no. 4, pp. 0416021–6, 2010.
18. M. Norgaard, O. Ravn, N. K. Poulsen, and L. K. Hansen, *Neural Networks for Modelling and Control of Dynamic Systems: A Practitioner's Handbook*, New York: Springer-Verlag Inc., 2003, 246 pp.
19. R. S. Burns, *Advanced Control Engineering*, Oxford, UK: Butterworth-Heinemann Publications, 2001.
20. M. Jelali and A. Kroll, *Hydraulic Servo-Systems: Modelling, Identification, and Control*, London: Springer-Verlag Ltd, 2004, 355 pp.
21. C. Junghui and H. Tien-Chih, Applying neural networks to on-line updated PID controllers for nonlinear process control, *Journal of Process Control*, vol. 14, no. 2, pp. 211–230, 2004.
22. P. Ailer, Comparison of linear and non-linear mathematical models developed for gas turbine control, in *Proceedings of 3rd International Conference on Nonlinear Problems in Aviation and Aerospace (ICNPAA)*, Daytona Beach, Florida, USA, May 10–12, 2000, pp. 11–19.
23. P. Ailer, Nonlinear mathematical modelling and control design developed for gas turbine, in *Proceedings of 7th Mini Conference on Vehicle System Dynamics, Identification and Anomalies (VSDIA)*, Budapest, Hungary, November 6–8, 2000, pp. 465–472.
24. P. Ailer, *Modelling and Nonlinear Control of a Low-Power Gas Turbine*, Department of Aircraft and Ships, Budapest University of Technology and Economics, 2001.
25. P. Ailer, G. Szederkényi, and K. M. Hangos, *Modeling and Nonlinear Analysis of a Low-Power Gas Turbine*, Hungary, 2001.
26. P. Ailer, G. Szederkényi, and K. M. Hangos, LPV-modeling of a low-power gas turbine, in *Proceedings of 8th Mini Conference on Vehicle System Dynamics, Identification and Anomalies (VSDIA)*, Budapest, Hungary, November 11–13, 2002, pp. 511–519.
27. P. Ailer, I. Santa, G. Szederkenyi, and K. M. Hangos, Nonlinear model-building of a low-power gas turbine, *Periodica Ploytechnica Transportation Engineering*, vol. 29, no. 1–2, pp. 117–135, 2002.
28. P. Ailer, G. Szederkényi and K. M. Hango, Parameter estimation and model validation of a low-power gas turbine, in *Proceedings of IASTED International Conference on Modelling, Identification and Control (MIC)*, Austria, February 18–21, 2002, pp. 604–609.
29. S. E. Abdollahi and A. Vahedi, *Dynamic Modelling of Micro-Turbine Generation Systems Using MATLAB/SIMULINK*, Tehran: Department of Electrical Engineering, Iran University of Science and Technology, 2004, 7 pp.
30. A. Aguiar, J. Pinto, and L. Nogueira, Modelling and simulation of natural gas micro-turbine for residential complexes, in *Proceedings of the World Congress on Engineering and Computer Science, WCECS*, San Francisco, USA, October 24–26, 2007, 4 pp.
31. G. Ofualagba, The modeling and simulation of a microturbine generation system, *International Journal of Scientific & Engineering Research*, vol. 2, no. 2, pp. 7 , 2012.

32. W. Rachtan and L. Malinowski, An approximate expression for part-load performance of a micro turbine combined heat and power system heat recovery unit, *Energy*, vol. 51, pp. 146–153, 2013.

33. L. Malinowski and M. Lewandowska, Analytical model-based energy and exergy analysis of a gas microturbine at part-load operation, *Applied Thermal Engineering*, vol. 57, no. 1–2, pp. 125–132, 2013.

34. A. G. Memon, K. Harijan, M. A. Uqaili, and R. A. Memon, Thermo-environmental and economic analysis of simple and regenerative gas turbine cycles with regression modeling and optimization, *Energy Conversion and Management*, vol. 76, pp. 852–864, 2013.

35. S. M. Hosseinalipour, E. Abdolahi, and M. Razaghi, Static and dynamic mathematical modeling of a micro gas turbine, *Journal of Mechanics*, vol. 29, no. 2, pp. 327–335, 2013.

36. W. I. Rowen, Simplified mathematical representations of heavy-duty gas turbines, *ASME Journal of Engineering for Power*, vol. 105, no. 4, pp. 865–869, 1983.

37. W. I. Rowen, Simplified mathematical representations of single-shaft gas turbines in mechanical drive service, *Turbomachinery International*, vol. 33, no. 5, pp. 26–32, 1992.

38. Y. S. H. Najjar, Performance of single-cycle gas turbine engines in two modes of operation, *Energy Conversion and Management*, vol. 35, no. 5, pp. 433–441, 1994.

39. R. Bettocchi, P. R. Spina, and F. Fabbri, *Dynamic Modeling of Single-Shaft Industrial Gas Turbine*, ASME 1996 International Gas Turbine and Aeroengine Congress and Exhibition, Birmingham, UK, June 10–13, 1996, vol. 4, pp. V004T11A007.

40. M. Bianchi, A. Peretto, and P. R. Spina, Modular dynamic model of multi-shaft gas turbine and validation test, in *Proceeding of The Winter Annual Meeting of ASME*, New York, 1998.

41. M. Mostafavi, A. Alaktiwi, and B. Agnew, Thermodynamic analysis of combined open-cycle twin-shaft gas turbine (Brayton Cycle) and exhaust gas operated absorption refrigeration unit, *Applied Thermal Engineering*, vol. 18, no. 9–10, pp. 847–856, 1998.

42. L. N. Hannett, G. Jee, and B. Fardanesh, A governor/turbine model for a twin-shaft combustion turbine, *IEEE Transactions on Power Systems*, vol. 10, no. 1, pp. 133–140, 1995.

43. B. E. Ricketts, Modelling of a gas turbine: A precursor to adaptive control, in *IEE Colloquium on Adaptive Controllers in Practice*, London, UK, 1997, pp. 7/1–7/5.

44. G. Crosa, F. Beltrami, A. Torelli, F. Pittaluga, A. Trucco, and F. Traverso, Heavy-duty gas turbine plant aerothermodynamic simulation using simulink, *ASME Journal of Engineering for Gas Turbines and Power*, vol. 120, no. 3, pp. 550–555, 1998.

45. M. Nagpal, A. Moshref, G. Morison, and P. Kundur, Experience with testing and modelling of gas turbines, in *Power Engineering Society Winter Meeting*, 2001, vol. 2, pp. 652–656.

46. J. Kaikko, T. Talonpoika, and P. Sarkomma, Gas turbine model for an on-line condition monitoring and diagnostic system, in *Australian Universities Power Engineering Conference (AUPEC)*, Melbourne, Australia, 2002, 7 pp.

47. Q. Z. Al-Hamdan and M. S. Y. Ebaid, Modeling and simulation of a gas turbine engine for power generation, *ASME Journal of Engineering for Gas Turbines and Power*, vol. 128, no. 2, pp. 302–311, 2006.

48. Y. Zhu and H. C. Frey, Simplified performance model of gas turbine combined-cycle systems, *Journal of Energy Engineering*, vol. 133, no. 2, pp. 82–90, 2007.
49. S. M. Camporeale, B. Fortunato, and M. Mastrovito, A modular code for real time dynamic simulation of gas turbines in SIMULINK, *ASME Journal of Engineering for Gas Turbines and Power*, vol. 128, no. 3, pp. 506–517, 2006.
50. H. Klang and A. Lindholm, *Modelling and Simulation of a Gas Turbine*, Norrkoping: Department of Science and Technology, Linkopings University, 2005.
51. J. Mantzaris and C. Vournas, Modelling and stability of a single-shaft combined-cycle power plant, *International Journal of Thermodynamics*, vol. 10, no. 2, pp. 71–78, 2007.
52. S. K. Yee, J. V. Milanović, and F. M. Hughes, Overview and comparative analysis of gas turbine models for system stability studies, *IEEE Transactions on Power Systems*, vol. 23, no. 1, pp. 108–118, 2008.
53. S. K. Yee, J. V. Milanović, and F. M. Hughes, Validated models for gas turbines based on thermodynamic relationships, *IEEE Transactions on Power Systems*, vol. 26, no. 1, pp. 270–281, 2011.
54. H. E. M. A. Shalan, M. A. Moustafa Hassan, and A. B. G. Bahgat, Comparative study on modelling of gas turbines in combined cycle power plants, in *Proceedings of the 14 th International Middle East Power Systems Conference (MEPCON)*, Cairo University, Egypt, December 19–21, 2010, pp. 970–976.
55. Q. C. Liang, W. Y. Lin, R. H. Huang, S. W. Jia, and J. M. Huang, A research on performance simulative models of gas turbines, *Kybernetes*, vol. 39, no. 6, pp. 1000–1008, 2010.
56. S. M. Hosseini, A. Fatehi, A. K. Sedigh, and T. A. Johansen, Automatic model bank selection in multiple model identification of gas turbine dynamics, *Journal of Systems and Control Engineering*, vol. 227, no. 5, pp. 482–494, 2013.
57. A. Lazzaretto and A. Toffolo, Prediction of performance and emissions of a two-shaft gas turbine from experimental data, *Applied Thermal Engineering*, vol. 28, no. 17–18, pp. 2405–2415, 2008.
58. N. M. Razali, *Development of an Analytical Model of Gas Turbine to Predict the Gas Turbine Degradation*, Tronoh, Perak: Universiti Teknologi Petronas, 2008.
59. M. Khosravi-el-hossani and Q. Dorosti, Improvement of gas turbine performance test in combine-cycle, *World Academy of Science, Engineering and Technology*, vol. 3, no. 10, pp. 377–380, 2009.
60. T. K. Ibrahim and M. M. Rahman, Effects of operation conditions on performance analysis of a gas turbine power plant, in *2nd National Conference in Mechanical Engineering Research and Postgraduate Studies (NCMER)*, Pahang, Malaysia, December 3–4, 2010, pp. 135–144.
61. T. K. Ibrahim and M. M. Rahman, Parametric study of a two-shaft gas turbine cycle model of power plant, in *1st International Conference on Mechanical Engineering Research 2011 (ICMER 2011)*, Malaysia, IOP Conf. Series: Materials Science and Engineering 36, 2012, 14 pp.
62. M. M. Rahman, T. K. Ibrahim, and A. Ahmed, Thermodynamic performance analysis of gas turbine power plant, *International Journal of the Physical Sciences*, vol. 6, no. 4, pp. 3539–3550, 2011.
63. M. M. Rahman, T. K. Ibrahim, M. Y. Taib, M. M. Noor, and R. A. Bakar, Thermal analysis of open-cycle regenerator gas-turbine power-plant, *World Academy of Science, Engineering and Technology*, vol. 44, pp. 801–806, 2010.

64. M. R. Bank Tavakoli, B. Vahidi, and W. Gawlik, An educational guide to extract the parameters of heavy-duty gas turbines model in dynamic studies based on operational data, *IEEE Transactions on Power Systems*, vol. 24, no. 3, pp. 1366–1374, 2009.

65. E. J. Roldan-Villasana, A. Vazquez, and V. M. Jimenez-Sanchez, Modelling of the simplified systems for a power plant simulator, in *Fourth UKSim European Symposium on Computer Modelling and Simulation (EMS)*, Pisa, Italy, November 17–19, 2010, pp. 277–282.

66. N. Yadav, I. A. Khan, and S. Grover, Modelling and analysis of simple open-cycle gas turbine using graph networks, *International Journal of Electrical and Electronics Engineering*, vol. 4, pp. 692–700, 2010.

67. P. T. Weber, *Modeling Gas Turbine Engine Performance at Part-Load*, Palo Alto, California: Electric Power Research Institute, Southwest Research Institute, University of Wyoming, 2011, 14 pp.

68. H. E. Shalan, M. M. A. Hassan, and A. B. G. Bahga, Parameter estimation and dynamic simulation of gas turbine model in combined-cycle power plants based on actual operational data, *Journal of American Science*, vol. 7, no. 5, pp. 303–310, 2011.

69. Y. Liu and M. Su, Nonlinear model based diagnostic of gas turbine faults: A case study, in *Proceedings of ASME Turbo Expo: Turbine Technical Conference and Exposition*, Vancouver, British Columbia, Canada, June 6–10, 2011, vol. 3.

70. J. Gao and Y. Huang, Effect of ambient temperature on three-shaft gas turbine performance under different control strategy, *Advanced Materials Research*, vol. 424–425, pp. 276–280, 2012.

71. E. Thirunavukarasu, *Modeling and Simulation Study of a Dynamic Gas Turbine System in a Virtual Test Bed*, South Carolina: College of Engineering and Computing, University of South Carolina, 2013.

72. P. Shaw, F. Zabihian, and A. S. Fung, Gas turbine-based combined cycle power plant modeling and effects of ambient temperature, in *7th International Conference on Energy Sustainability*, Minneapolis, Minnesota, USA, July 14–19, 2013, pp. V001T02A002.

73. A. P. Wiese, M. J. Blom, M. J. Brear, C. Manzie, and A. Kitchener, Development and validation of a physics-based, dynamic model of a gas turbine, in *ASME Turbo Expo: Turbine Technical Conference and Exposition*, San Antonio, Texas, USA, June 3, 2013, vol. 4, 10 pp.

74. M. M. A. Al-Sood, K. K. Matrawy, and Y. M. Abdel-Rahim, Optimum parametric performance characterization of an irreversible gas turbine brayton cycle, *International Journal of Energy and Environmental Engineering*, vol. 4, no. 37, 2013.

75. S.-K. Kim, P. Pilidis, and J. Yin, Gas turbine dynamic simulation using SIMULINK, in Society of Automotive Engineers (SAE) Technical Papers, 2000, pp. 231–237.

76. C. Evans, D. Rees, and D. Hill, Frequency domain identification of gas turbine dynamics, *IEEE Transactions on Control Systems Technology*, vol. 6, no. 5, pp. 651–662, 1998.

77. C. Evans, N. Chiras, P. Guillaume, and D. Rees, *Multivariable Modelling of Gas Turbine Dynamics*, University of Glamorgan & Vrije Universiteit Brussel, 2001.

78. V. Arkov, C. Evans, P. J. Fleming, D. C. Hill, J. P. Norton, I. Pratt, D. Rees, and K. Rodrlguez-Vfizquez, System identification strategies applied to aircraft gas turbine engines, *Annual Reviews in Control*, vol. 24, no. 1, pp. 67–81, 2000.

79. V. Arkov, G. Kulikov, and T. Breikin, Life cycle support for dynamic modelling of gas turbines, in *15th Triennial World Congress*, Barcelona, Spain, 2002, vol. 15, pp. 913–918.

80. C. Riegler, M. Bauer, and J. Kurzke, Some aspects of modeling compressor behaviori in gas turbine performance calculations, *ASME Journal of Turbomachinery*, vol. 123, no. 2, pp. 372–378, 2000.

81. A. Behbahani, R. K. Yedavalli, P. Shankar, and M. Siddiqi, Modeling, diagnostics and prognostics of a two-spool turbofan engine, in *Proceedings of the 41st AIAA/ASME/SAE/ASEE Joint Propulsion Conference and Exhibit (AIAA)*, Tucson, Ariz, USA, July 2005, 14 pp.

82. H. Pourfarzaneh, A. Hajilouy-Benisi, and M. Farshch, An analytical model of a gas turbine components performance and its experimental validation, in *ASME Turbo Expo: Power for Land, Sea, and Air*, Glasgow, UK, June 14–18, 2010, vol. 1, pp. 335–340.

83. G.-Y. Chung, M. Dhingra, J. V. R. Prasad, M. Richard, and S. Sirica, An analytical approach to gas turbine engine model linearization, in *ASME Turbo Expo: Turbine Technical Conference and Exposition*, Vancouver, British Columbia, Canada, June 6–10, 2011, vol. 3, pp. 105–115.

84. Z. Abbasfard, *Fault Diagnosis of Gas Turbine Engines by Using Multiple Model Approach*, Montreal, Quebec: Concordia University, 2013.

85. F. Lu, Y. Chen, J. Huang, D. Zhang, and N. Liu, An integrated nonlinear model-based approach to gas turbine engine sensor fault diagnostics, in *Proceedings of the Institution of Mechanical Engineers, Part G: Journal of Aerospace Engineering*, vol. 228, no. 11, pp. 2007–2021, 2013.

86. F. Lu, Y. Lv, J. Huang, and X. Qiu, A model-based approach for gas turbine engine performance optimal estimation, *Asian Journal of Control*, vol. 15, no. 6, pp. 1794–1808, 2013.

87. P. Ailer, Mathematical modelling of control system of a low-power engine, in *Proceedings of "The Challenge of Next Millennium on Hungarian Aeronautical Sciences" Conference*, Budapest, Hungary, June 2–4, 1999, pp. 142–152.

88. P. Ailer, G. Szederkényi, and K. M. Hangos, Nonlinear control design of a low-power gas turbine, in *Proceedings of 2nd International PhD Workshop on Systems and Control; a Young Generation Viewpoint*, Balatonfüred, Hungary, September 17–20, 2001, pp. 1–10.

89. P. Ailer, G. Szederkényi, and K. M. Hangos, Model-based nonlinear control of a low-power gas turbine, in *Proceedings of 15th Triennial World Congress of the International Federation of Automatic Control*, Barcelona, Spain, 2002, vol. 15, pp. 222–227.

90. P. Ailer, G. Szederkényi, and K. M. Hangos, LPV-analysis and control design of a low power gas turbine, in *Proceedings of 4th International PhD Workshop on Information Technologies and Control; a Young Generation Viewpoint*, Libverda, Czech Republic, September 16–20, 2003.

91. P. Ailer, B. Pongrácz, and G. Szederkényi, Constrained control of a low power industrial gas turbine based on input-output linearization, in *Proceedings of International Conference on Control and Automation (ICCA)*, Budapest, Hungary, 2005, pp. 368–373.

92. J. L. Agüero, M. C. Beroqui, and H. D. Pasquo, *Gas Turbine Control Modifications for: Availability and Limitation of Spinning Reserve and Limitation of Non-Desired Unloading*, Pluspetrol Energy SA, Tucumán, Argentina, 2002, 8 pp.

93. P. Centeno, I. Egido, C. Domingo, F. Fernández, L. Rouco, and M. González, *Review of Gas Turbine Models for Power System Stability Studies*, Madrid: Universidad Pontificia Comillas and Endesa Generación, 2002.

94. M. Ashikaga, Y. Kohno, M. Higashi, K. Nagai, and M. Ryu, A study on applying nonlinear control to gas turbine systems, in *International Gas Turbine Congress (IGTC)*, Tokyo, Japan, November 2–7, 2003, 8 pp.

95. C. Zaiet, O. Akhrif, and L. Saydy, Modeling and non linear control of a gas turbine, in *IEEE International Symposium on Industrial Electronics*, Montreal, Quebec, Canada, July 9–13, 2006, pp. 2688–2694.

96. M. Lichtsinder and Y. Levy, Jet engine model for control and real-time simulations, *ASME Journal of Engineering for Gas Turbines and Power*, vol. 128, no. 4, pp. 745–753, 2006.

97. B. Pongraz, P. Ailer, K. M. Hangos, and G. Szederkenyi, Nonlinear reference tracking control of a gas turbine with load torque estimation, *International Journal of Adaptive Control and Signal Processing*, vol. 22, no. 8, pp. 757–773, 2008.

98. J. P. Tong and T. Yu, Nonlinear PID control design for improving stability of micro-turbine systems, in *Third International Conference on Electric Utility Deregulation and Restructuring and Power Technologies (DRPT)*, Nanjuing, China, April 6–9, 2008, pp. 2515–2518.

99. E. Najimi and M. H. Ramezani, Robust control of speed and temperature in a power plant gas turbine, *ISA Transactions*, vol. 51, no. 2, pp. 304–308, 2012.

100. I. V. Kolmanovsky, L. C. Jaw, W. Merrill, and H. T. Van, Robust control and limit protection in aircraft gas turbine engines, in *IEEE International Conference on Control Applications (CCA)*, Dubrovnik, October 3–5, 2012, pp. 812–819.

101. M. Pakmehr, *Towards Verifiable Adaptive Control of Gas Turbine Engines*, Georgia: School of Aerospace Engineering, Georgia Institute of Technology, 2013.

102. S. J. Qin and T. A. Badgwell, *An Overview of Industrial Model Predictive Control Technology*, University of Texas at Austin, USA, 1997, 31 pp.

103. S. Qin and T. A. Badgwell, A survey of industrial model predictive control technology, *Control Engineering Practice*, vol. 11, no. 7, pp. 733–764, 2003.

104. J. Richalet, A. Rault, J. L. Testud, and J. Papon, Model predictive heuristic control: Application to industrial processess, *Automatica*, vol. 14, no. 5, pp. 413–428, 1978.

105. J. Richalet, Industrial applications of model based predictive control, *Automatica*, vol. 29, no. 5, pp. 1251–1274, 1993.

106. M. Nikolaou, Model predictive controllers: A critical synthesis of theory and industrial needs, *Advances in Chemical Engineering*, vol. 26, pp. 131–204, 2001.

107. J. B. Rawlings, Tutorial overview of model predictive control, *Control Systems*, vol. 20, no. 3, pp. 38–52, 2000.

108. R. K. Agrawal and M. Yunis, A Generalized mathematical model to estimate gas turbine starting characteristics, *ASME Journal of Engineering for Gas Turbines and Power*, vol. 104, no. 1, pp. 194–201, 1982.

109. S. R. Balakrishnan and S. Santhakumar, Fuzzy modeling considerations in an aero gas turbine engine start cycle, *Fuzzy Sets and Systems*, vol. 78, no. 1, pp. 1–4, 1996.

110. A. Peretto and P. R. Spina, Comparison of industrial gas turbine transient responses performed by different dynamic models, in *Turbo Expo*, Orlando, Florida, USA, June 2–5, 1997, 9 pp.

111. T. B. Henricks, *An Investigation into Computer Simulation of the Dynamic Response of a Gas Turbine Engine,* Department of Mechanical Engineering, The Graduate School, The Pennsylvania State University, Pennsylvania, USA, 1997, 52 pp.

112. A. Beyene and T. Fredlund, *Comparative Analysis of Gas Turbine Engine Starting,* ASME 1998 International Gas Turbine and Aeroengine Congress and Exhibition, Stockholm, Sweden, June 2–5, 1998, vol. 4, pp. V004T11A009.

113. J. H. Kim, T. W. Song, T. S. Kim, and S. T. Ro, Dynamic simulation of full startup procedure of heavy-duty gas turbines, *ASME Journal of Engineering for Gas Turbines and Power,* vol. 124, no. 3, pp. 510–516, 2002.

114. J. H. Kim, T. W. Song, T. S. Kim, and S. T. Ro, Model development and simulation of transient behavior of heavy duty gas turbines, *ASME Journal of Engineering for Gas Turbines and Power,* vol. 123, no. 3, pp. 589–594, 2000.

115. T. S. Kim, H. J. Park, and S. T. Ro, Characteristics of transient operation of a dual-pressure bottoming system for the combined cycle power plant, *Energy,* vol. 26, no. 10, pp. 905–918, 2001.

116. J. Y. Shin, Y. J. Jeon, D. J. Maeng, J. S. Kim, and S. T. Ro, Analysis of the dynamic characteristics of a combined-cycle power plant, *Energy,* vol. 27, no. 12, pp. 1085–1098, 2002.

117. C. R. Davison and A. M. Birk, Steady state and transient modeling of a micro-turbine with comparison to operating engine, in *ASME Turbo Expo: Power for Land, Sea, and Air,* Vienna, Austria, June 14–17, 2004, pp. 27–35.

118. X. H. Huang and X. S. Zheng, Research on startup model of aircraft engine based on stage-stacking method, *Acta Aeronautica et Astronautica Sinica,* vol. 26, no. 5, pp. 524–528, 2005.

119. W. Xunkai and L. Yinghong, Aero-engine dynamic start model based on parsimonious genetic programming, in *6th World Congress on Intelligent Control and Automation (WCICA),* Dalian, January 2006, pp. 1478–1482.

120. S. Sanaye and M. Rezazadeh, Transient thermal modeling of heat recovery steam generators in combined cycle power plants, *International Journal of Energy Research,* vol. 31, no. 11, pp. 1047–1063, 2007.

121. G. Kocer, *Aerothermodynamic Modelling and Simulation of Gas Turbines for Transient Operating Conditions,* Aerospace Engineering Department, Middle East Technical University, Turkey, 2008, 74 pp.

122. M. Corbett, P. Lamm, P. Owen, S. D. Phillips, M. Blackwelder, J. T. Alt, J. McNichols, M. Boyd, and M. Wolff, Transient turbine engine modeling with hardware-in-the-loop power extraction, in *6th International Energy Conversion Engineering Conference (IECEC),* Cleveland, Ohio, July 2008, 7 pp.

123. F. Alobaid, R. Postler, J. Strohle, B. Epple, and K. Hyun-Gee, Modeling and investigation start-up procedures of a combined cycle power plant, *Applied Energy,* vol. 85, no. 12, pp. 1173–1189, 2008.

124. X. Y. Zhang, Nonlinear model-based predictor: Its application to the closed loop control of the alstom GT11N2 gas turbine, in *ASME Turbo Expo: Turbine Technical Conference and Exposition,* Vancouver, British Columbia, Canada, June 6–10, 2011, vol. 3, pp. 243–250.

125. R. Rezvani, M. Ozcan, B. Kestner, J. Tai, D. N. Mavris, R. Meisner, and S. Sirica, A gas turbine engine model of transient operation across the flight envelope, in *ASME Turbo Expo: Turbine Technical Conference and Exposition,* Vancouver, British Columbia, Canada, January 1, 2011, pp. 133–140.

126. K. Daneshvar, A. Behbahani-nia, Y. Khazraii, and A. Ghaedi, Transient modeling of single-pressure combined cycle power plant exposed to load reduction, *International Journal of Modeling and Optimization (IJMO)*, vol. 2, no. 1, pp. 64–70, 2012.

127. M. Rahnama, H. Ghorbani, and A. Montazeri, Nonlinear identification of a gas turbine system in transient operation mode using neural network, in *4th Conference on Thermal Power Plants (CTPP)*, Tehran, Iran, December 18–19, 2012, pp. 1–6.

128. M. H. Refan, S. H. Taghavi, and A. Afshar, Identification of heavy duty gas turbine startup mode by neural networks, in *4th Conference on Thermal Power Plants (CTPP)*, Tehran, Iran, December 18–19, 2012, pp. 1–6.

129. S. Sarkar, K. Mukherjee, S. Sarkar, and A. Ray, Symbolic transient time-series analysis for fault detection in aircraft gas turbine engines, in *Proceedings of the American Control Conference*, Montreal, QC, June 27–29, 2012, pp. 5132–5137.

130. M. Venturini, Simulation of compressor transient behavior through recurrent neural network models, *ASME Journal of Turbomachinery*, vol. 128, no. 3, pp. 444–454, 2005.

131. M. Venturini, Optimization of a real-time simulator based on recurrent neural networks for compressor transient behavior prediction, *ASME Journal of Turbomachinery*, vol. 129, no. 3, pp. 468–478, 2006.

132. M. Venturini, Development and experimental validation of a compressor dynamic model, *ASME Journal of Turbomachinery*, vol. 127, no. 3, pp. 599–608, 2004.

133. M. Morini, M. Pinelli, and M. Venturini, Development of a one-dimensional modular dynamic model for the simulation of surge in compression systems, *ASME Journal of Turbomachinery*, vol. 129, no. 3, pp. 437–447, 2006.

134. M. Morini, M. Pinelli, and M. Venturini, Analysis of biogas compression system dynamics, *Applied Energy*, vol. 86, no. 11, pp. 2466–2475, 2009.

135. D. Henrion, L. Reberga, J. Bernussou, and F. Vary, *Linearization and Identification of Aircraft Turbofan Engine*, LAAS-CNRS and SNECMA Moteurs, France: IFAC, 2004.

136. F. Jurado, Non-linear modeling of micro-turbines using NARX structures on the distribution feeder, *Energy Conversion and Management*, vol. 46, no. 3, pp. 385–401, 2005.

137. C. Bartolini, F. Caresana, G. Comodi, L. Pelag, M. Renzi, and S. Vagni, Application of artificial neural networks to micro gas turbines, *Energy Conversion and Management*, vol. 52, no. 1, pp. 781–788, 2011.

138. H. Nikpey, M. Assadi and P. Breuhaus, Development of an optimized artificial neural network model for combined heat and power micro gas turbines, *Applied Energy*, vol. 108, pp. 137–148, 2013.

139. A. Lazzaretto and A. Toffolo, Analytical and neural network models for gas turbine design and off-design simulation, *International Journal of Applied Thermodynamics*, vol. 4, no. 4, pp. 173–182, 2001.

140. S. Ogaji, R. Singh and S. Probert, Multiple-sensor fault-diagnoses for a two-shaft stationary gas turbine, *Applied Energy*, vol. 71, no. 4, pp. 321–339, 2002.

141. J. Arriagada, M. Genrup, A. Loberg and M. Assadi, Fault diagnosis system for an industrial gas turbine by means of neural networks, in *International Gas Turbine Congress*, Tokyo, November 2–7, 2003, 6 pp.

142. M. Basso, L. Giarre, S. Groppi and G. Zappa, NARX models of an industrial power plant gas turbine, *IEEE Transactions on Control Systems Technology*, vol. 13, no. 4, pp. 599–604, 2005.

143. M. Kaiadi, *Artificial Neural Networks Modelling for Monitoring and Performance Analysis of a Heat and Power Plant*, Lund: Division of Thermal Power Engineering, Department of Energy Sciences, Faculty of Engineering, Lund University, 2006.

144. R. Bettocchi, M. Pinelli, P. R. Spina and M. Burgio, Set up of a robust neural network for gas turbine simulation, in *ASME Turbo Expo*, Vienna, Austria, June 14–17, 2004, pp. 543–551.

145. R. Bettocchi, M. Pinelli, P. Spina and M. Venturini, Artificial intelligence for the diagnostics of gas turbines—Part I: Neural network approach, *ASME Journal of Engineering for Gas Turbines and Power*, vol. 129, no. 3, pp. 711–719, 2007.

146. P. Spina and M. Venturini, Gas turbine modelling by using neural networks trained on field operating data, in *ECOS*, Padova, Italy, June 25–28, 2007, 29 pp.

147. S. Simani and R. Patton, Fault diagnosis of an industrial gas turbine prototype using a system identification approach, *Control Engineering Practice*, vol. 16, no. 7, pp. 769–786, 2008.

148. Y. Yoru, T. H. Karakoc and A. Hepbasli, Application of artificial neural network (ANN) method to exergetic analyses of gas turbines, in *International Symposium on Heat Transfer in Gas Turbine Systems*, Antalya, Turkey, August 9–14, 2009, 4 pp.

149. M. Fast, M. Assadi and D. Sudipta, Condition based maintenance of gas turbines using simulation data and artificial neural network: A demonstration of feasibility, in *ASME Turbo Expo*, Berlin, Germany, June 9–13, 2008, 9 pp.

150. M. Fast, M. Assadi, and S. De, Development and multi-utility of an ANN model for an industrial gas turbine, *Journal of Applied Energy*, vol. 86, no. 1, pp. 9–17, 2009.

151. M. Fast, T. Palmé, and M. Genrup, A novel approach for gas turbines monitoring combining CUSUM technique and artificial neural network, in *ASME Turbo Expo*, Orlando, Florida, USA, June 8–12, 2009, 8 pp.

152. M. Fast and T. Palmé, Application of artificial neural network to the condition monitoring and diagnosis of a combined heat and power plant, *Energy*, vol. 35, no. 2, pp. 1114–1120, 2010.

153. M. Fast, *Artificial Neural Networks for Gas Turbine Monitoring*, Lund: Division of Thermal Power Engineering, Department of Energy Sciences, Faculty of Engineering, Lund University, 2010.

154. T. Palmé, M. Fast, and M. Thern, Gas turbine sensor validation through classification with artificial neural networks, *Applied Energy*, vol. 88, no. 11, pp. 3898–3904, 2011.

155. H. A. Nozari, H. D. Banadaki, M. A. Shoorehdeli, and S. Simani, Model-based fault detection and isolation using neural networks: An industrial gas turbine case study, in *21st International Conference on Systems Engineering (ICSEng)*, Las Vegas, NV, USA, August 16–18, 2011, pp. 26–31.

156. H. A. Nozaria, M. A. Shoorehdeli, S. Simani, and H. D. Banadaki, Model-based robust fault detection and isolation of an industrial gas turbine prototype using soft computing techniques, *Neurocomputing*, vol. 91, no. 8, pp. 29–47, 2012.

157. N. Chiras, C. Evans, and D. Rees, Nonlinear gas turbine modelling using NARMAX structures, *IEEE Transactions on Instrumentation and Measurement*, vol. 50, no. 4, pp. 893–898, 2001.

158. N. Chiras, C. Evans, and D. Rees, Nonlinear modelling and validation of an aircraft gas turbine engine, in *5th IFAC Symposium on Nonlinear Control Systems*, ST Petersburg, Russia, July 4–6, 2001, vols 1–3, pp. 871–876.

159. N. Chiras, C. Evans, and D. Rees, *Nonlinear Gas Turbine Modelling Using Feedforward Neural Networks*, Wales: School of Electronics, University of Glamorgan, 2002.

160. N. Chiras, C. Evans, and D. Rees, Global nonlinear modeling of gas turbine dynamics using NARMAX structures, *ASME Journal of Engineering for Gas Turbines and Power*, vol. 124, no. 4, pp. 817–826, 2002.

161. A. E. Ruano, P. J. Fleming, C. Teixeira, K. Rodriguez-Vazquez, and C. M. Fonseca, Nonlinear identification of aircraft gas-turbine dynamics, *Neurocomputing*, vol. 55, no. 3–4, pp. 551–579, 2003.

162. G. Torella, F. Gamma, and G. Palmesano, Neural networks for the study of gas turbine engines air system, in *International Gas Turbine Congress*, Tokyo, Japan, November 2–7, 2003, 8 pp.

163. T. V. Breikin, G. G. Kulikov, P. J. F. Arkov, and P. J. Fleming, Dynamic modelling for condition monitoring of gas turbines: Genetic algorithms approach, in *16th IFAC World Congress*, Czech Republic, 2005, vol. 16, pp. 123–123.

164. H. Badihi, A. Shahriari, and A. Naghsh, Artificial neural network application to fuel flow function for demanded jet engine performance, in *Proceedings of IEEE Aerospace Conference*, Mont, USA, March 7–14, 2009, pp. 1–7.

165. R. Mohammadi, E. Naderi, K. Khorasani, and S. Hashtrudi-Zad, Fault diagnosis of gas turbine engines by using dynamic neural networks, in *IEEE International Conference on Quality and Reliability (ICQR)*, Bangkok, Thailand, September 14–17, 2011, pp. 25–30.

166. I. Loboda, Y. Feldshteyn, and V. Pon, Neural networks for gas turbine fault identification: Multilayer perceptron or radial basis network?, *International Journal of Turbo and Jet-Engines*, vol. 29, no. 1, pp. 37–48, 2012.

167. S. S. Tayarani-Bathaie, Z. Vanini, and K. Khorasani, Fault detection of gas turbine engines using dynamic neural networks, in *25th IEEE Canadian Conference on Electrical & Computer Engineering (CCECE)*, Montreal, QC, April 29–May 2, 2012, pp. 1–5.

168. M. Kulyk, S. Dmitriev, O. Yakushenko, and O. Popov, Method of formulating input parameters of neural network for diagnosing gas-turbine engines, *Aviation*, vol. 17, no. 2, pp. 52–56, 2013.

169. M. Agarwal, A systematic classification of neural-network-based control, *Control Systems*, vol. 17, no. 2, pp. 75–93, 1997.

170. A. M. Suarez, M. A. Duarte-Mermoud, and D. F. Bassi, A predictive control scheme based on neural networks, *Kybernetes*, vol. 35, no. 10, pp. 1579–1606, 2006.

171. A. Draeger, S. Engell, and H. Ranke, *Model Predictive Control Using Neural Networks*, 1995, pp. 61–66.

172. M. Ö. Efe, Neural network-based control, in Bogdan M. Wilamowski, and J. David Irwin (eds.), *Intelligent Systems*, Boca Raton, Florida, USA: Taylor and Francis Group, LLC, 2011, 26 pp.

173. K. J. Hunt, D. Sbarbaro, R. Żbikowski, and P. J. Gawthrop, Neural networks for control systems—A survey, *Automatica*, vol. 28, no. 6, pp. 1083–1112, 1992.

174. S. Balakrishnan and R. D. Weil, Neurocontrol: A literature survey, *Mathematical and Computer Modelling*, vol. 23, no. 1–2, pp. 101–117, 1996.

175. W. I. Rowen and R. L. Van Housen, Gas turbine airflow control for optimum heat recovery, *ASME Journal of Engineering for Power*, vol. 105, pp. 72–79, 1983.

176. M. T. Hagan, H. B. Demuth, and O. D. JESÚS, An introduction to the use of Neural Networks in Control Systems, *International Journal of Robust and Nonlinear Control*, vol. 12, no. 11, pp. 959–985, 2002.

177. M. T. Hagan and H. B. Demuth, Neural networks for control, in *American Control Conference*, San Diego, CA, USA, June 1999, pp. 1642–1656.

178. I. T. Nabney and D. C. Cressy, Neural network control of a gas turbine, *Neural Computing & Applications*, vol. 4, no. 4, pp. 198–208, 1996.

179. N. Dodd and J. Martin, Using neural networks to optimize gas turbine aero engines, *Computing & Control Engineering Journal*, vol. 8, no. 3, pp. 129–135, 1997.

180. D. Psaltis, A. Sideris, and A. A. Yamamura, A multilayered neural network controller, *IEEE Control Systems Magazine*, 1988, pp. 17–21.

181. K. Lietzau and A. Kreiner, Model based control concepts for jet engines, in *Turbo Expo 2001*, New Orleans, Louisiana, USA, June 4–7, 2001, 8 pp.

182. K. Lietzau and A. Kreiner, *The Use of Onboard Real-Time Models for Jet Engine Control*, Germany: MTU Aero Engine, 2004.

183. P. K. Dash, S. Mishra, and G. Panda, A radial basis function neural network controller for UPFC, *IEEE Transactions on Power Systems*, vol. 15, no. 4, pp. 1293–1299, 2000.

184. Y. Becerikli, A. Konar, and S. Tarıq, Intelligent optimal control with dynamic neural networks, *Neural Networks,* vol. 16, no. 2, pp. 251–259, 2003.

185. J. S. Litt, K. I. Parker, and S. Chatterjee, Adaptive gas turbine engine control for deterioration compensation due to aging, in *16th International Symposium on Airbreathing Engines*, Cleveland, Ohio, October 2003, pp. 212607/1–212607/8.

186. G. Lalor and M. O'Malley, Frequency control on an Island power system with increasing proportions of combined cycle gas turbines, in *Proceedings of Power Tech Conference (IEEE)*, Bolognia, Italy, June 23–26, 2003, vol. 4, 7 pp.

187. S. Sahin and A. Savran, A neural network approach to model predictive control, in *14th Turkish Symposium on Artificial Intelligence and Neural Networks (TAINN)*, Turkey, June 16–17, 2005, pp. 386–392.

188. M. Ławryńczuk, A family of model predictive control algorithms with artificial neural networks, *International Journal of Applied Mathematics and Computer Science*, vol. 17, no. 2, pp. 217–232, 2007, pp. 777–184.

189. M. Ławrynczuk, An efficient nonlinear predictive control algorithm with neural models based on multipoint on-line linearisation, in *International Conference on "Computer as a Tool" (EUROCON)*, Warsaw, September 9–12, 2007.

190. A. Jadlovská, N. Kabakov, and J. Sarnovský, Predictive control design based on neural model of a non-linear system, *Acta Polytechnica Hungarica*, vol. 5, pp. 93–108, 2008.

191. A. Z. Cipriano, Fuzzy predictive control for power plants, in *Advanced Fuzzy Logic Technologies in Industrial Applications*, 2006, pp. 279–297.

192. J. S. Kim, K. M. Powell, and T. F. Edgar, Nonlinear model predictive control for a heavy-duty gas turbine power plant, in *American Control Conference (ACC)*, Washington DC, June 17–19, 2013 pp. 2952–2957.

193. H. Ghorbani, A. Ghaffari, and M. Rahnama, Constrained model predictive control implementation for a heavy-duty gas turbine power plant, *WSEAS Transactions on Systems and Control*, vol. 3, no. 6, pp. 507–516, 2008.

194. H. Ghorbani, A. Ghaffari, and M. Rahnama, Multivariable model predictive control for a gas turbine power plant, in *10th WSEAS International Conference on Automatic Control, Modelling & Simulation (ACMOS)*, Istanbul, Turkey, May 27–30, 2008, pp. 275–181.

195. J. Mu and D. D. Rees, Approximate model predictive control for gas turbine engines, in *American Control Conference*, Boston, Massachusetts, June 30–July 2, 2004, pp. 5704–5709.

196. J. X. Mu, D. Rees, and G. P. Liu, Advanced controller design for aircraft gas turbine engines, *Control Engineering Practice*, vol. 13, no. 8, pp. 1001–1015, 2005.

197. A. M. Schaefer, D. Schneegass, V. Sterzing, and S. Udluft, A neural reinforcement learning approach to gas turbine control, in *International Joint Conference on Neural Networks (IJCNN)*, Orlando, Florida, USA, August 12–17, 2007, pp. 1691–1696.

198. N. Sisworahardjo, M. El-Sharkh, and M. Alam, Neural network controller for microturbine power plants, *Electric Power Systems Research*, vol. 78, no. 8, pp. 1378–1384, 2008.

199. J. Yamagami, K. Okajima, O. Koyama, S. Yamamoto, and H. Oonuki, Development of next generation gas turbine control systems, *IHI Engineering Review*, vol. 41, no. 2, pp. 74–79, 2008.

200. M. Bazazzadeh, H. Badihi, and A. Shahriari, Gas turbine engine control design using fuzzy logic and neural networks, *International Journal of Aerospace Engineering*, 12 pp. 2011.

201. S. Balamurugan, R. J. Xavier, and A. Jeyakumar, ANN controller for heavy-duty gas turbine plant, *International Journal of Applied Engineering Research*, vol. 3, no. 12, pp. 1765–1771, 2008.

202. R. Bettocchi, M. Pinelli, P. R. Spina, M. Venturini, and G. A. Zanetta, Assessment of the robustness of gas turbine diagnostics tools based on neural networks, in *ASME Turbo Expo: Power for Land, Sea, and Air*, Barcelona, Spain, May 8–11, 2006, vo. 4, pp. 603–613.

203. R. Bettocchi, M. Pinelli, P. R. Spina, and M. Venturini, Artificial intelligence for the diagnostics of gas turbines. Part II: Neuro-fuzzy approach, *ASME Journal of Engineering for Gas Turbines and Power,* vol. 129, no. 3, pp. 720–729, 2006.

204. K. Ghorbani and M. Gholamrezaei, Axial compressor performance map prediction using artificial neural network, in *Proceedings of ASME Turbo Expo*, Montreal, Canada, May 14–17, 2007, vol. 6, pp. 1199–1208.

205. K. Ghorbani and M. Gholamrezaei, An artificial neural network approach to compressor performance prediction, *Applied Energy,* vol. 86, no. 7–8, pp. 1210–1221, 2009.

206. T. Palmé, P. Waniczek, H. Hönen, M. Assadi, and P. Jeschke, Compressor map prediction by neural networks, *Journal of Energy & Power Engineering*, vol. 6, no. 10, pp. 1651, 2012.

207. A. A. Mozafari and M. Lahroodi, Modeling and control of gas turbine combustor with dynamic and adaptive neural networks, *International Journal of Engineering (IJE), Transactions B: Applications*, vol. 21, no. 1, pp. 71–84, 2008.

208. V. Sethi, G. Doulgeris, P. Pilidis, A. Nind, M. Doussinault, P. Cobas, and A. Rueda, The map fitting tool methodology: Gas turbine compressor off-design performance modeling, *ASME Journal of Turbomachinery*, vol. 135, no. 6, pp. 15, 2013.

209. H. Asgari, X. Q. Chen, and R. Sainudiin, Applications of artificial neural networks to rotating equipment, in *3rd Conference on Rotating Equipment in Oil and Power Industries*, Tehran, Iran, January 2012, 10 pp.

210. M. Caudill, Neural network primer: Part I, *AI Expert*, vol. 2, no. 12, pp. 46–52, 1989.

211. M. T. Hagan, H. B. Demuth, and M. H. Beale, *Neural Network Design, Boulder, Colorado*: Campus Publishing Service, University of Colorado at Boulder, 2002, 736 pp.

212. M. H. Beale, M. T. Hagan, and H. B. Demuth, Neural Network Toolbox™ User's Guide, R2011b ed., Natick, MA: MathWorks, 2011, 404 pp.

213. P. S. Curtiss and D. D. Massie, Neural Network Fundamentals for Scientists and Engineers, 2001. [Online]. Available: www.cibse.org/pdfs/neural.pdf.

214. G. Cybenko, Approximation by superpositions of a sigmoidal function, *Mathematics of Control, Signals, and Systems*, vol. 2, no. 4, pp. 303–314, 1989.

215. B. Karlik and A. V. Olgac, Performance analysis of various activation functions in generalized MLP architectures of neural networks, *International Journal of Artificial Intelligence and Expert Systems (IJAE)*, vol. 1, no. 4, pp. 111–122, 2010.

216. K. Debes, A. Koenig, and H. M. Gross, *Transfer Functions in Artificial Neural Networks; A Simulation-Based Tutorial*, Ilmenau: Department of Neuroinformatics and Cognitive Robotics, Technical University Ilmenau, 11 pp.

217. G. W. Irwin, K. Warwick, and K. J. Hunt, *Neural Network Applications in Control*, vol. 53, Institution of Electrical Engineers: IEE Control Engineering Series, London, UK, 1995, pp. 295.

218. H. Asgari, X. Q. Chen, and R. Sainudiin, Analysis of ANN-based modelling approach for industrial applications, *International Journal of Innovation, Management and Technology (IJIMT)*, vol. 4, no. 1, pp. 165–169, 2013.

219. H. Asgari, X. Q. Chen, and R. Sainudiin, Analysis of ANN-based modelling approach for industrial applications, in *International Conference on Industrial Applications and Innovations (ICIAI)*, Hong Kong, February 2013, pp. 165–169.

220. J. Bourquin, H. Schmidli, P. V. Hoogevest, and H. Leuenberger, Advantages of artificial neural networks (ANNs) as alternative modelling technique for data sets showing non-linear relationships using data from a galenical study on a solid dosage form, *European Journal of Pharmaceutical Sciences*, vol. 7, no. 1, pp. 5–16, 1998.

221. J. Bourquin, H. Schmidli, P. V. Hoogevest, and H. Leuenberger, Comparison of artificial neural networks (ANN) with classical modelling techniques using different experimental designs and data from a galenical study on a solid dosage form, *European Journal of Pharmaceutical Sciences*, vol. 6, no. 4, pp. 287–301, 1998.

222. M. T. Haque and A. M. Kashtiban, Application of neural networks in power systems; A review, in *Proceedings of World Academy of Science, Engineering and Technology*, Boston, MA, June 6, 2005, pp. 53–57.

223. G. E. Hinton, How neural networks learn from experience, in *Cognitive Modeling*, vol. 267, MIT Press, 2002, pp. 181–195.

224. H. White, Learning in artificial neural networks: A statistical perspective, *Neural Computation*, vol. 1, no. 4, pp. 425–464, 1989.

225. A. Guez and J. Selinsky, A neuromorphic controller with a human teacher, in *IEEE International Conference on Neural Networks*, San Diego, CA, USA, July 1988, vol. 2, pp. 595–602.

226. J. V. Tu, Advantages and disadvantages of using artificial neural networks versus logistic regression for predicting medical outcomes, *Journal of Clinical Epidemiology*, vol. 49, no. 11, pp. 1225–1231, 1996.

227. H. Asgari, X. Q. Chen, M. B. Menhaj, and R. Sainudiin, ANN-based system identification, modelling, and control of gas turbines—A review, in *International Conference on Power and Energy Engineering (ICPEE)*, Phuket Island, Thailand, 2012, 7 pp.

228. H. Asgari, X. Q. Chen, M. B. Menhaj, and R. Sainudiin, ANN-based system identification, modelling, and control of gas turbines—A review, *Manufacturing Science and Technology III, Advanced Materials Research*, vol. 622–623, pp. 611–617, 2013.

229. H. Asgari, X. Q. Chen, M. B. Menhaj, and R. Sainudiin, Artificial neural network–based system identification for a single-shaft gas turbine, *ASME Journal of Engineering for Gas Turbines and Power*, vol. 135, no. 9, pp. 092601–7, 2013.

230. H. Asgari, X. Q. Chen, R. Sainudiin, M. Morini, M. Pinelli, P. R. Spina, and M. Venturini, Modeling and simulation of the startup operation of a heavy-duty gas turbine using NARX models, in *Turbo Expo*, Düsseldorf, Germany, June 16–20, 2014, 10 pp.

231. H. Asgari, M. Venturini, X. Q. Chen, and R. Sainudiin, Modelling and simulation of the transient behavior of an industrial power plant gas turbine, *ASME Journal of Engineering for Gas Turbines and Power*, vol. 136, no. 6, pp. 061601, 10 pages, 2013.

232. M. Morini, G. Cataldi, M. Pinelli, and M. Venturini, A model for the simulation of large-size single-shaft gas turbine start-up based on operating data fitting, in *ASME Turbo Expo*, Montreal, Canada, May 14–17, 2007, 23 pp.

233. P. P. Walsh and P. Fletcher, *Gas Turbine Performance*, 2nd ed., Tulsa: Pennwell Books, 1998, 656 pp.

234. C. Cutler and B. Ramaker, Dynamic matrix control—A computer control algorithm, in *Joint Automation Control Conference*, San Francisco, California, 1980.

235. P. Orukpe, *Basics of Model Predictive Control*, Imperial College, London, 2005, 27 pp.

236. D. Soloway and P. J. Haley, Neural generalized predictive control, in *IEEE International Symposium on Intelligent Control*, Dearborn, MI, USA, September 15–18, 1996, pp. 277–282.

237. K. S. Narendra and S. Mukhopadhyay, Adaptive control using neural networks and approximate models, *IEEE Transactions on Neural Networks*, vol. 8, no. 3, pp. 475–485, 1997.

238. M. Araki, PID control, in H. Unbehauen (ed.), *Control, Systems, Robotics, and Automation*, vol. 2, Luiz Carlos da Silva, Kyoto University, Kyoto, Japan, 1996, http://www.scribd.com.

239. K. J. Åström, PID control, in *Control System Design*, Santa Barbara, CA, 2002, pp. 216–251.

240. Wikimedia Commons 2013. [Online]. Available: http://commons.wikimedia.org. [Accessed 2013].

241. D. Viswanath, Study of Minimum and Non-Minimum Phase Systems, 2009. [Online]. Available: http://www.scribd.com/doc/16067103/study-of-nonminimum-phase-systems. [Accessed January 10, 2014].

242. C. R. Holsonback, *Dynamic Thermal-Mechanical-Electrical Modeling of the Integrated Power System of a Notional All-Electric Naval Surface Ship*, Austin, Texas: The University of Texas at Austin, 2007.

# Index